WITHDRAWN

SCIENCE, DECISION AND VALUE

THE UNIVERSITY OF WESTERN ONTARIO
SERIES IN PHILOSOPHY OF SCIENCE

A SERIES OF BOOKS
ON PHILOSOPHY OF SCIENCE, METHODOLOGY,
AND EPISTEMOLOGY
PUBLISHED IN CONNECTION WITH
THE UNIVERSITY OF WESTERN ONTARIO
PHILOSOPHY OF SCIENCE PROGRAMME

Managing Editor

J. J. LEACH

Editorial Board

J. BUB, R. E. BUTTS, W. HARPER, J. HINTIKKA, D. J. HOCKNEY,

C. A. HOOKER, J. NICHOLAS, G. PEARCE

VOLUME 1

SCIENCE, DECISION AND VALUE

PROCEEDINGS OF THE FIFTH UNIVERSITY OF
WESTERN ONTARIO PHILOSOPHY COLLOQUIUM, 1969

Edited by

JAMES LEACH, ROBERT BUTTS, AND GLENN PEARCE

University of Western Ontario, Ontario, Canada

D. REIDEL PUBLISHING COMPANY

DORDRECHT-HOLLAND/BOSTON-U.S.A.

Library of Congress Catalog Card Number 72-77877

ISBN 90 277 0327 2

Published by D. Reidel Publishing Company,
P.O. Box 17, Dordrecht, Holland

Sold and distributed in the U.S.A., Canada, and Mexico
by D. Reidel Publishing Company, Inc.
306 Dartmouth Street, Boston,
Mass. 02116, U.S.A.

BF
441
U55
1969

All Rights Reserved
Copyright © 1973 by D. Reidel Publishing Company, Dordrecht, Holland
No part of this book may be reproduced in any form, by print, photoprint, microfilm,
or any other means, without written permission from the publisher

Printed in The Netherlands by D. Reidel, Dordrecht

PREFACE

This volume grew out of the papers and comments presented at the Fifth University of Western Ontario Philosophy Colloquium, October 31–November 2, 1969. The colloquium papers were delivered by P. Suppes, R. B. Braithwaite, C. W. Churchman, and J. S. Minas. Comments are provided from others attending the colloquium, with one reply by P. Suppes.

Also included are papers recently published elsewhere by A. Michalos, P. Fishburn and H.-N. Castañeda. The editors express thanks to these authors and to the editors of the following respective journals for permission to publish: *Theory and Decision*, *Synthese*, and *Critica*. Finally, there is an extensive bibliography of decision theory, vis-à-vis science and values.

The editors wish to thank the officers of the University of Western Ontario for making the colloquium possible.

THE EDITORS

CONTENTS

PREFACE	V
PATRICK SUPPES / The Concept of Obligation in the Context of Decision Theory	1
HENRY KYBURG / Comments	15
PATRICK SUPPES / Reply to Professor Kyburg	19
R. B. BRAITHWAITE / Behind Decision and Games Theory: Acting with a Co-Agent versus Acting Along with Nature	22
ISAAC LEVI / Comments	56
RONALD GIERE / Comments	62
I. J. GOOD / Comments	67
C. WEST CHURCHMAN / Measurement: A Systems Approach	70
ISAAC LEVI / Comments	87
RONALD GIERE / Comments	95
PETER C. FISHBURN / Utility Theory with Inexact Preferences and Degrees of Preference	98
I. J. GOOD / Information, Rewards, and Quasi-Utilities	115
HECTOR-NERI CASTAÑEDA / Open Action, Utility, and Utilitarianism	128
ROBERT BINKLEY / Comments	148
J. S. MINAS / Emergent Utilities	156
ALEX C. MICHALOS / Cost-Benefit versus Expected Utility Acceptance Rules	163
ALAN ROSS and DANNY STEINBERG / Bibliography	191

PATRICK SUPPES

THE CONCEPT OF OBLIGATION IN THE CONTEXT OF DECISION THEORY

I. INTRODUCTION

Three observations have been the stimulus to my thinking about obligation. One is the penetrating but baffling paper by H. A. Prichard (1932) entitled 'Duty and Ignorance of Fact.' I agree with almost everything Prichard has to say, at least insofar as I understand it, but I always finish rereading the essay with a feeling of incompleteness. What can be said about duty in the face of partial ignorance? Prichard does not make the next move to partial degrees of belief and the outlines of decision theory.

The second observation is that there seems to be no natural way to introduce the concept of obligation into a complete decision theory, fully equipped with a numerical subjective probability function and a numerical utility function, and the single rule of behavior always to maximize expected utility. The concept of obligation must be put in some early stage, almost surely at the point of extending the primitive at concepts of decision theory.

The third observation is that obligatory acts stand to acts in general as certain events stand to events in general – and here by *certain events* I mean those events that must occur with certainty. If A is a certain event and $A \subseteq B$ then B also is a certain event. The event that must occur is X, the set of all possible outcomes. If $A \subseteq X$ but $A \neq X$ then A may be subjectively certain but is not logically necessary. Here 'logically necessary' means that the occurrence of such an event A is imposed by the structure of the Boolean algebra of events, and not by the subjective probability measure or qualitative relation defined on the algebra. My inference from what has just been said about events and the analogy that constitutes the third observation is that we need to make explicit the logical structure of obligation in just the same formal way that we make explicit the logical structure of certainty in subjective probability theory and decision theory.

The rest of what I have to say shall be addressed to this task. I am sure I shall say a number of things that are wrong, but I look forward to learning from your corrections of my intuitions. I do not know that the theory I work out here can be used to answer any questions of real significance, but developing it has at least clarified my own thinking about obligation.

II. PRIMITIVE CONCEPTS OF THE THEORY

Because I want to make as explicit as possible the independence of the theory of obligation from the theory of utility, I shall deliberately leave out some of the concepts of standard decision theory – especially the concept of the set C of consequences, on which the utility function is ordinarily defined.

I retain X, the set of states of nature, and a set D of decisions defined on X. From a formal standpoint the elements of D are functions whose domains of definition are X. We could of course now introduce the set of consequences C as the union of the ranges of the functions in D, but I deliberately refrain from so doing.

Ordinarily in decision theory we would next consider a preference ordering on D, and then by appropriate definition – using some such device as constant decisions or functions – extend it to a probability ordering on the events that are a subset of X.

For our purposes here it is good enough just to assume both these orderings, and this we may do with one primitive, an ordering $\geqslant$ that is a subset of $(D \times D) \cup (X \times X)$. We shall when needed use $>$ for strict preference and $\approx$ for indifference, with their standard definitions in terms of $\geqslant$ being understood. On D we talk about a preference ordering, perhaps one generated by obligations; on X, we talk about a subjective probability ordering. The axioms given below relate the two.

Finally, we come to the last primitive, the one that is especially important here. This is the set $\mathcal{O}$ of obligatory acts. However, $\mathcal{O}$ is not simply a subset of D, the full given set of acts or decisions, but rather a subset of conditional or partial acts. The idea is this. Ordinarily in decision theory an act or decision is defined on all states of nature and represents a decision as to how to act under all possible states of affairs. Such decisions are total functions, in a familiar terminology. Often only parts of such functions would be considered obligatory. Consider this example. Let A be the event of a two-year-old child's starting to run across a busy street in Jones' presence. Let f be the act or decision that for every x in A is the act of stopping the child, and let f, for $x \notin A$, be the (passive) act of watching the child play. Then intuitively, not f, but f restricted to A, in symbols: $f \mid A$, is the obligatory act. So $\mathcal{O}$ is a set of such partial functions, i.e., $\mathcal{O}$ is just the set of obligatory acts.

The theory of obligation developed here is then stated in terms of a structure $\chi = \langle X, D, \mathcal{O}, \geqslant \rangle$, where $X, D, \mathcal{O}$ and $\geqslant$ have the set-theoretical properties described above – X and D must be nonempty but not necessarily $\mathcal{O}$; and as events we take all subsets of X.

III. AXIOMS OF THE THEORY

The two most important ideas expressed in the axioms are the closure properties making explicit how new obligatory acts may be formed out of given ones, and the preference ordering of obligatory acts, as well as the ordering of such acts in relation to other acts. In addition, problems of consistency among obligations must also be faced.

Further remarks about the axioms and the statement of alternatives will be easier to formulate with the axioms already in front of us. I do note that de Finetti's qualitative axioms for subjective belief or probability are included because of the importance I attach to making explicit the relation between obligation and ignorance of fact, endorsing in the process Prichard's subjective view of the matter.

We also need one formal definition for the preference or choice axiom. This is the notion of the *maximum obligatory domain* of a decision or act, abbreviated MOD, and defined as follows:

A is the MOD *of f if and only if* $f \mid A \in \mathcal{O}$ *and if* $A \subseteq B$, *but* $A \neq B$ *then* $f \mid B \notin \mathcal{O}$.

DEFINITION. *A structure* $\chi = \langle X, D, \mathcal{O}, \geqslant \rangle$ *is a structure of obligation if and only if the following axioms are satisfied for all events A, B and C and all decisions f and g.*

Belief Axioms

B1. *The relation* $\geqslant$ *is a weak ordering on X.*
B2. $A \geqslant \emptyset$.

B3.   Not $\emptyset \geqslant X$.

B4.   If $A \cap C = B \cap C = \emptyset$ then $A \geqslant B$ if and only if $A \cup C \geqslant B \cup C$.

Closure Axioms

Cl1.   If $A \subseteq B$ and $f \,|\, B \in \mathcal{O}$ then $f \,|\, A \in \mathcal{O}$.

Cl2.   If $f \,|\, A, g \,|\, B \in \mathcal{O}$ and $A \cap B = \emptyset$ then $f \,|\, A \cup g \,|\, B \in \mathcal{O}$.

Consistency Axiom

Con.   If $f \,|\, A, g \,|\, B \in \mathcal{O}$ and $A \cap B \neq \emptyset$ then $f \,|\, A \cap B = g \,|\, A \cap B$.

Preference or Choice Axiom

Ch.   If A is the MOD of f and B is the MOD of g, and either $A > \emptyset$ or $B > \emptyset$ then $f \geqslant g$ if and only if $A \geqslant B$.

Because the belief axioms have been commented on extensively in the literature of decision theory I shall not say anything about them here. The two closure axioms do not guarantee that the set $\mathcal{O}$ of obligatory acts is nonempty, for in many decision situations it is reasonable to suppose that there are no obligatory acts, and the concept of obligation does not have direct relevance to a choice among acts. The first closure axiom does say that if a partial function or partial act is obligatory then a further restriction to a smaller domain, that is, a restriction to a more restricted and more specific event will also be obligatory. To extend the example given earlier, if B is the event of a two-year-old child's starting to run into a busy street and A is the more specific event of the child's being two feet from the curb as well, then the closure axiom requires that given it is obligatory to stop the child in the case of event B it is also obligatory in the case of event A. The second closure axiom specifies a method of building up complex obligatory acts from simpler ones. If f is obligatory given event A and g is obligatory given event B, and if the intersection of A and B is empty so that not both A and B can occur together, then the union of the two obligatory acts is also obligatory.

The consistency axiom is more controversial than the preceding axioms in terms of standard discussions of obligation. It is sometimes maintained that obligations can be inconsistent, but the function of this axiom is that in a given structure of obligations no such inconsistency can occur. If two acts are obligatory and their domains overlap then the axiom requires that the two acts be identical on the common domain, that is, on the common event $A \cap B$.

By far the strongest axiom is the last one, the preference or choice axiom. This axiom asserts that one act will be preferred to another or chosen in place of another just on the basis of obligation and without any consideration of utility. This means that whenever there are any acts that are obligatory and the events on which they are conditioned have nonnull probability, then obligation dominates all other considerations of utility. Utility or desirability would enter only in choosing between two obligatory acts whose conditional events had equal probability of occurring. The second aspect of this axiom, which represents a strong assumption, is that the only basis for choice among obligatory acts is the probability of the conditional events' occurring. This part of the axiom answers in a very strong form the question left unanswered by Prichard. Roughly speaking, the axiom asserts that apart from information all obligatory acts are equivalent in terms of preference or choice; one act is not per se more obligatory than another, given that both belong to the set of obligatory acts. The only basis for choice is information. An obligatory act conditioned on the more probable event should be the one chosen. To paraphrase the title of Prichard's article, the force of this axiom is to say that duty depends on factual belief. To put it more strongly, we might even say that the force of the axiom is that all duties are equally obligatory, and the choice among them depends *only* on factual beliefs.

In all likelihood the choice axiom is too strong and needs to be weakened in any one of several ways. The point of stating it here is to make explicit what might be called the simple theory of obligation. What is not clear to me is how to use something other than degree of factual belief or desirability as a basis for arbitrating between various obligations. Certainly we all recognize that one obligation can be overridden by another. For example, it is an obligation not to shove other people while walking on the sidewalk, but that obligation can be overridden by the obligation of saving a child from being hit by a car. In order to reach the child in time, it would be considered appropriate by almost everyone to rudely push another adult. The obligation to be polite and considerate while walking on a sidewalk is relatively minor compared to the obligation to make an effort to prevent harm to a child. The present choice axiom clearly does not differentiate between these two obligations; it could even be that the act of shoving the adult is a more certain violation of obligation than the act of reaching for the child. What seems to be needed is an additional structure of preference or choice on obligations independent of considerations of information. This additional structure should depend upon the seriousness or importance of the obligations. Even then, I foresee difficulties, for if obligatory act f given A is more serious or important than obligatory act g given B, it may still be the case that we should not select f over g because the probability of event A's occurring is very small in comparison to the probability of B's occurring. In other words, it will not be sufficient simply to introduce an independent hierarchy of seriousness. However, it is clear that a natural apparatus can be drawn from decision theory to take account of this hierarchical problem, namely, we can think of assigning a weight of seriousness to each obligatory act, and then compute expected seriousness by taking expectation with respect to the subjective factual beliefs. I shall not work out the details of this additional development here, but it constitutes a natural extension of the simpler theory that is the central focus of the present paper.

IV. SOME ELEMENTARY THEOREMS

It is easy to prove a number of elementary theorems about obligation which follow from the axioms. Some examples are given here. The proofs will mostly be omitted.

THEOREM 1. *If f_A, $g_B \in \mathcal{O}$ then $f_A \cup g_B \in \mathcal{O}$.*

This theorem asserts an unrestricted closure property of $\mathcal{O}$ that is stronger than Axiom Cl 2, but follows from this axiom and the consistency axiom.

THEOREM 2. *If A is the MOD of f and B is the MOD of g, and $A \subseteq B$ then $f \leqslant g$, provided either $A > \emptyset$ or $B > \emptyset$.*

This theorem says that if whenever A occurs B must occur and if A is the maximum obligatory domain of f, and B of g, then g should be (weakly) preferred to f.

THEOREM 3. *If $A \cap B = \emptyset$, A is the MOD of f, B is the MOD of g, and $B > \emptyset$, then any function h in D such that $h \mid A \cup B = f \mid A \cup g \mid B$ is preferred to f, i.e., $h > f$.*

This theorem gives a method for building up preferred 'composite' obligatory acts.

THEOREM 4. *Let F be the set of all acts f such that the* MOD *of f is strictly more probable than* ∅. *Then F is weakly ordered by* ⩾, *i.e.,* ⩾ *is transitive and strongly connected on F.*

V. COMPARISON WITH DEONTIC LOGIC

It will be instructive to compare some of the systematic characteristics of obligatory acts in the present theory with the view of obligatory acts developed in deontic logic.

One of the first things noticeable is that the concept of negation does not naturally apply in the present theory to acts as functions. It is especially not natural to talk about the negation of an act. From an uninteresting set-theoretical standpoint the complement of the act considered as a function is defined with respect to a universe that can be specified but the concept of complementation put in this fashion does not have an interesting interpretation, for the complement of an act is not an act. What should be explicitly noted is that the ordinary sentential language and the ordinary use of the sentential connectives do not apply in a simple way to acts considered as functions.

A detailed comparison with deontic logic does deepen the feeling for the structure of the present theory. In his classical article, von Wright (1951) gives a number of laws of deontic logic. Von Wright's notation is this. Propositions are denoted by initial capital letters of the alphabet; OA means that A is obligatory and PA that A is permitted.

To begin with, there is the tautology

$$PA \leftrightarrow \neg(O \neg A),$$

where $\neg$ means negation and $\leftrightarrow$ means *if and only if*. Translation into our setup requires definition of a permitted (partial) act. Let us call the set of such acts $\mathscr{P}$, which we define as follows.

DEFINITION. $f \mid A \in \mathscr{P}$ *if and only if there is no g and no event* $B \subseteq A$ *such that* $g \mid B \in \mathcal{O}$ *and* $f \mid B \neq g \mid B$.

The conceptual similarity to von Wright's equivalence is apparent. A more complex point to raise in the context of decision theory is this. We might want to liberalize the notion of permitted acts by requiring that the event B have positive probability, i.e., $B > \emptyset$, or in familiar terminology of probability theory we ignore sets of measure zero.

Von Wright's second law is that OA entails PA. Here and in what follows I shall write the corresponding property in the present theory as a theorem, numbering consecutively with the theorem of the last section.

THEOREM 5. *If $f \mid A \in \mathcal{O}$ then $f \mid A \in \mathcal{P}$.*

This theorem follows essentially immediately from the consistency axiom and the definition of $\mathcal{P}$.

Next follow four laws that von Wright calls laws for the dissolution of operators.

a. $O(A \text{ \& } B)$ is identical with $(OA) \text{ \& } (OB)$,
b. $P(A \vee B)$ is identical with $(PA) \vee (PB)$,
c. $(OA) \vee (OB)$ entails $O(A \vee B)$
d. $P(A \text{ \& } B)$ entails $(PA) \text{ \& } (PB)$.

Corresponding to (a) we have:

THEOREM 6. *If $f \mid A, f \mid B \in \mathcal{O}$ then $f \mid A \cap B \in \mathcal{O}$.*

On the other hand, it does not follow, as in the case of 'half' of (a) that if $f \mid A \cap B \in \mathcal{O}$ then $f \mid A \in \mathcal{O}$ and $f \mid B \in \mathcal{O}$. So here a clear conceptual difference exists.

Corresponding to (b), we have:

THEOREM 7. *If $f \mid A \cup g \mid B$ is a function, then $f \mid A \cup g \mid B \in \mathcal{P}$ if and only if $f \mid A, g \mid B \in \mathcal{P}$.*

Again, we miss a full analogue, which would be

$$f \mid A \cup g \mid B \in \mathcal{P} \text{ if and only if } f \mid A \in \mathcal{P} \text{ or } g \mid B \in \mathcal{P};$$

here the implication from right to left is false.

Corresponding to (c) we have an even poorer analogue, but a complete analogue of Theorem 7, with obligation replacing permission.

THEOREM 8.

$$f \mid A \cup g \mid B \in \mathcal{O} \text{ if and only if } f \mid A, g \mid B \in \mathcal{O}.$$

In the case of Theorem 8 we do not need the hypothesis of Theorem 7, for Theorem 1 guarantees that the union of any two obligatory acts is also obligatory, a part of our strong requirement of consistency.

As to (d), we have:

THEOREM 9. *If $f \mid A, g \mid B \in \mathscr{P}$ then $f \mid A \cap g \mid B \in \mathscr{P}$, but not conversely.*

Disparities between deontic logic and the present theory are already evident. They are made even more so when we consider von Wright's six laws of commitment. The first one asserts that $OA \& O(A \to B)$ entails OB. The absence of negation of an act as function has already been noted. There are similar difficulties with an adequate notion of implication, for the Boolean operation corresponding to implication is ordinarily defined in terms of complementation and union, or complementation and intersection.

In view of these problems, it is natural to seek for an approach within the present theory that is set-theoretic rather than function-theoretic in character. This we can do along the following lines. First, rather than use von Wright's notation, because of confusion with the event notation used here, let us use lower-case letters 'p', 'q', 'r', etc., for acts, and let an act now be a *set* of partial functions in the sense of this paper. To avoid overlapping terminology let us call these new acts k-acts, where the 'k' stands for kind, since acts as functions represent more closely individual acts rather than kinds of acts. Second, it will be simplest to relativize k-acts to a given event or state of affairs A. Generally the reference to A shall be omitted, and A shall be assumed constant. Granted the relativization the acts are now total. In other words, p is a k-act (relative to A) if and only if p is a subset of function acts $f \mid A$. For formal explicitness, we define

$$\mathscr{A}(A) = \{f \mid A : f \in D\}.$$

Then

DEFINITION. *p is a k-act (relative to A) if and only if p is a subset of $\mathscr{A}$.*
Complementation or negation is defined relative to $\mathscr{A}$, and implication is given its usual Boolean definition. We define Op to mean that p is an obligatory k-act as follows.

DEFINITION. *Op (relative to A) if and only if there is a partial function $f \mid A$ in p and $f \mid A \in \mathcal{O}$.*

It follows, of course, from our axioms that there is at most one such $f \mid A \in \mathcal{O}$, i.e., essentially from the consistency axiom, there can be no conflict of obligation relative to a given state of affairs A.

From this point on we can follow von Wright rather closely. First we define permission.

Pp if and only if not $O\neg p$.

The following theorems then follow easily, using mainly the Consistency Axiom. The first four correspond to (a)–(d) above.

THEOREM 10. *$O(p \cap q)$ if and only if Op and Oq.*
THEOREM 11. *$P(p \cup q)$ if and only if Pp or Pq.*
Proof: By virtue of Theorem 10, we have

$O(\neg p \cap \neg q)$ if and only if $O\neg p$ and $O\neg q$,

whence by elementary logic

not $O(\neg p \cap \neg q)$ if and only if not $(O\neg p$ and $O\neg q)$,

and thus by applying de Morgan's law to the Boolean algebra on the left and to the sentential logic on the right, we have

not $O\neg(p \cup q)$ if and only if not $O\neg p$ or not $O\neg q$.

Using then the definition of permission, we obtain the theorem.

THEOREM 12. *Op or Oq if and only if $O(p \cup q)$.*

Proof: Going from left to right, by hypothesis there is an $f\,|\,A$ in $\mathcal{O}$ such that either $f\,|\,A$ is in p or $f\,|\,A$ is in q, whence $f\,|\,A$ is in $p \cup q$. The converse is similar.

In Theorem 12 we find a remaining fundamental difference from von Wright's logic. In his system, Op or Oq entails $O(p \cup q)$, but not conversely, for he has in mind that a disjunction of acts may be obligatory without either member being so. That is not the case here because of the generation of obligatory k-acts from a given state of affairs A and at most one function-theoretic obligatory act.

THEOREM 13. *If $P(p \cap q)$ then Pp and Pq.*

In the case of this theorem, we match exactly the strength of von Wright's logic. The converse implication holds in neither system. The absence of this converse is closely related to another question discussed by von Wright. He asks what should be the logical status of the propositions $P(p \cap \neg p)$ and $O(p \cup \neg p)$. It might be thought both should be theorems – the empty k-act is permitted and the universal k-act is obligatory, but von Wright suggests the best course is to regard them as expressing contingent propositions, and with this view the present theory is in agreement. In order for $O(p \cup \neg p)$ to hold, for instance, an obligatory partial act $f \mid A$ must be in $\mathcal{O}$, but this is not required. Indeed, it would be contrary to the spirit of the present theory to require that for each $A > \emptyset$ there be an $f \mid A$ in $\mathcal{O}$; only an ultra-Calvinist view of the world could tolerate such a feature.

I now turn to von Wright's six laws of commitments, formulated as Theorems 14, 15, 17–20. I note explicitly that Boolean implication $p \to q$ is defined as $\neg p \cup q$.

THEOREM 14. *If Op and $O(p \to q)$ then Oq.*

Proof: By hypothesis $f \mid A$ is in p, in $\neg p \cup q$, and also in $\mathcal{O}$, and so by the Boolean version of tollendo ponens, $f \mid A$ is in q.

THEOREM 15. *If Pp and $O(p \to q)$ then Pq.*

Proof: By hypothesis there is an $f \mid A$ in $\neg p \cup q$ and also in $\mathcal{O}$, but by the definition of P, $f \mid A$ is not in $\neg p$, and so $f \mid A$ is in q, and thus Pq.

However, as the last statement of the proof shows, we can strengthen this theorem to the conclusion that q is obligatory. This conclusion runs counter to von Wright's intended interpretation, and is certainly a controversial feature of the present setup. Von Wright had in mind something like the following. Promising is permitted. It is obligatory if a promise is given to keep it. Consequently keeping promises is permitted. In this instance a temporal sequence of acts constitutes the intended interpretation, but in the underlying model of the present theory the situation is static. Given the state of affairs A, exactly one partial act $f \mid A$ can be performed, and it may or may not be obligatory. What this adds up to is this. If there is an obligatory k-act p, then for any k-act q if q is permitted it is also obligatory. I formulate this principle as a theorem to be scrutinized more intensely later.

THEOREM 16. *If Op and Pq then Oq.*

I now return to von Wright's third law of commitment, stated as Theorem 17.

THEOREM 17. *If not Pq and $O(p \to q)$ then not Pp.*

Proof: By hypothesis $f \mid A$ is in $\neg p \cup q$, in $\neg q$ and in $\mathcal{O}$, whence $f \mid A$ is in $\neg p$, and thus not Pp.

THEOREM 18. *If $O(p \to (q \cup r))$, not Pq and not Pr then not Pp.*

Proof: By hypothesis $f \mid A$ is in $\neg p \cup q \cup r$ and in $\mathcal{O}$, but $f \mid A$ is not in q and not in r. So $f \mid A$ is in $\neg p$, and thus not Pp.

THEOREM 19. *If Op and $O((p \cap q) \to r)$ then $O(q \to r)$.*

Proof: By hypothesis $f \mid A$ is in p and in $\mathcal{O}$, and also in $p \cap q \to r$, but $p \cap q \to r = p \to (q \to r)$, so $f \mid A$ is in $q \to r$, as desired.

THEOREM 20. *If $O(\neg p \to p)$ then Op.*

Proof: Follows at once from the Boolean identity $\neg p \to p = p$.

From the theorems that have just been proved it is clear that the obligatory theory of k-acts is somewhat stronger than von Wright's deontic logic. In particular, Theorems 12 and 16 are not true in his theory.

However, Theorem 12 fails – it just becomes an entailment rather than equivalence – if we broaden the definition of k-act so that it is not restricted to a given event A. To avoid confusion with the definition of k-acts already given, I use 'α', 'β', 'γ', etc., to stand for these broadened k-acts. I shall still refer to these new objects as k-acts but the notation will signal that the new definition is being used. First, we define $\mathcal{F}$ as the set of all partial functions. Formally,

$$\mathcal{F} = \{f \mid A : A \subseteq X \text{ and } f \in D\}.$$

Then

DEFINITION. *α is a k-act if and only if α is a subset of $\mathcal{F}$.*

The only other change is to relativize obligation to a given state of affairs A. In other words, in the modified theory of k-acts, obligation rather than the k-acts themselves is relativized to A. Intuitively this fits well the general view of this paper that obligations do not hold absolutely but relative to factual beliefs about the true state of affairs. The formal definition of Op (relative to A) can stand as given earlier, but now it is $O\alpha$, and α itself is not restricted to A.

Using this wider concept of k-act we may look at Theorem 12 and see why now we may not infer from $O(\alpha \cup \beta)$ that $O\alpha$ or $O\beta$. As a counterexample let $f \mid A$ be in $\mathcal{O}$, not in α and not in β, but let $A = B \cup C$ with $f \mid B$ in α and $f \mid C$ in β.

As to Theorem 16, it is not changed by the broadened concept of k-act. Suppose α is obligatory. Then $f \mid A$ is in α and also in $\mathcal{O}$. Now suppose β is permitted. Then $f \mid A$ is not in $\neg \beta$, but since $f \mid A$ is in α, it is in $\beta \cup \neg \beta$, and thus in β, so β is obligatory.

Because Theorem 16 holds also under the modified concept of k-act, it is worth scrutinizing its meaning more carefully. As a truth about the general theory of moral obligation it seems too hard a saying even for those of Calvinistic bent. It is considerably more reasonable when it is assessed in the context of the decision context used here. The theorem applies when we are given a state of affairs A and there is an obligatory act α in circumstances A. In this environment any permissible act β must be *identical with* α, and hence obligatory. Put another way, in circumstances that mandate an obligatory act, any other act is forbidden.

Because of the clarity and explicitness of von Wright's classic article on deontic logic it has been a natural touchstone for analyzing the theory of obligation from a decision-theoretic viewpoint. However, in his book, *Norm and Action* (1963) von Wright expresses dissatisfaction with several aspects of the 1951 article. The principal ones are these: (i) the definition of permission in terms of obligation and negation, as given above; (ii) the principles of distributivity and commitment discussed above; (iii) the treatment of acts as propositions and the consequent application of sentential connectives to form new, complex acts.

The last point I have already dwelled upon. In a decision-theoretic context it is natural – and by now almost a convention – to treat acts as functions, not as propositions or events. The problem of permission and its definability in terms of obligation is especially severe in the theory I have developed, in view of Theorem 16. But I do emphasize that Theorem 16 does *not* hold for individual acts taken as partial functions. In other words, there are acts $f \mid A$ and $g \mid B$ such that $f \mid A \in \mathcal{O}$, $g \mid B \in \mathcal{P}$ and $g \mid B \notin \mathcal{O}$. On the other hand, we do have an analogue of Theorem 16 when the state of affairs A is held constant.

THEOREM 21. *If* $f \mid A \in \mathcal{O}$ *and* $g \mid A \in \mathcal{P}$ *then* $g \mid A \in \mathcal{O}$.

The theorem follows from the fact that in the theory postulated here, for a given A there can be just one obligatory act, and therefore $g \mid A = f \mid A$.

Concerning the principles of distributivity and commitment stated in von Wright's 1951 article, I shall not expand further upon the comments made already when their formal status in the present theory was examined.

VI. FURTHER AXIOMS

Comparison with deontic logic is useful because of the effort that has been made to be clear about the logical structure of the deontic modalities, but moral philosophy will remain weak in substantive structure and content if it remains formally at the broad level of generality of deontic logic. Richer structural assumptions are needed to make deeper contact with the concept of obligation that has been the focus of so much philosophical analysis.

Working out the concept of expected seriousness has already been alluded to. The hard part of this effort, developing a hierarchy of seriousness, or, as we might say, a hierarchy of obligations, is as yet mainly untouched. It is hardly likely that intuitive considerations alone will be sufficient to construct such a hierarchy, and it should be obvious from earlier remarks that I would scarcely expect to use a priori arguments to justify the necessarily detailed additional axioms that are needed.

The spirit of investigation should be that which dominates normative economics – the detailed analysis of alternative formal structures. In the theory of moral obligation, as in the theory of most other concepts in moral philosophy, there has as yet been too much Lockean clearing of the underbrush and too little Newtonian building. The present paper is meant to be a modest contribution to the Newtonian task.

Stanford University

BIBLIOGRAPHY

Prichard, H. A., 'Duty and Ignorance of Fact', in *Moral Obligation, Essays and Lectures*, Clarendon Press, Oxford, 1949, p. 18–39.
von Wright, G. H., 'Deontic Logic', *Mind* **60** (1951), 1–15.
von Wright, G. H., *Norm and Action*, Routledge and Kegan Paul, London, 1963.

HENRY KYBURG

COMMENTS

I have little to say of a critical nature about Professor Suppes' paper; it is, like so much else that he has done, a gem of sparkling clarity. It is one more boulder in the mounting mountain of evidence that set theory has rendered many of the other techniques of philosophical inquiry obsolete. One of the virtues of the set theoretic approach is that it disarms criticism. If the set theoretical philosopher has done his work without making mistakes, the only thing for a set theoretical commentator to do is to generalize or extend his results. Set theoretical philosophers do make mistakes, of course. But I doubt that Professor Suppes has, and so I have not spent my time looking for contradictions. Nor have I attempted to extend and generalize the theorems of his paper. I have instead (somewhat reluctantly) abandoned the role of set theoretical commentator in favor of the older role of conversational critic. I hope that the few items I have decided to mention will at any rate serve to stimulate discussion.

Professor Suppes draws mainly on two philosophical works: an article by H. A. Prichard (1932), in which the problem of the relation between obligations and factual knowledge is discussed, and a paper by G. H. von Wright from *Mind*, 1951 called 'Deontic Logic'. The paper by Prichard is a British talk-piece. (You know the style: the important point is the question of the plausibility or implausibility of doing a thing *A*, or refraining from doing *A*, or doing not *A*, in the face of knowledge of *B* or ignorance of *B*, except when *B* or knowledge of *B* entails *A* or knowledge of *A*.... I find it very hard to read those things.) At any rate, there is not much of substance there. The paper by von Wright is a short and clear statement of some very plausible and very simple principles of deontic logic. I asked myself why *these* two papers?

Prichard is explicable – he was perhaps the first to state the problem clearly. But von Wright's 1951 paper is neither the latest nor the most thorough examination of deontic logic. Von Wright himself wrote a long treatise on the subject, *Norm and Action*, which appeared in 1963. He has a new article on deontic logic and its relations with modal logic in a recent issue of *Critica*. And of course, others have worked on deontic logic. Why only refer to the earlier article?

One answer is that Suppes' system parallels the 1951 von Wright system in a clear and interesting way. The complexities that von Wright introduces in *Norm and Action*, the modal considerations introduced in the *Critica* article, simply don't fit in to the set theoretical structure that Professor Suppes offers.

This fact is not without significance. The significance is not that Professor Suppes is lacking in ingenuity, perseverance, or energy – we all know better than that. It is that Professor Suppes is concerned about a very different thing than that which bothered von Wright or, for that matter, Prichard. I think von Wright and the others have been concerned with the logical structure of statements or propositions concerning obligation. They hope in this way not only to explicate the structure of deontic argument, but also to gain some insight into the nature of obligation itself. Professor Suppes, on the other hand, is concerned with the problem of the application of the concept of obligation in a decision theoretic framework. He is trying to exhibit the structure underlying the *application* of this concept.

The two tasks are not unrelated, but they are two tasks. That there are parallels between the theorems of von Wright's elementary 1951 deontic logic and Suppes' 1969 set theoretical decision theory is interesting, but it doesn't go very far; and it certainly neither amplifies nor clarifies what von Wright and others have done since then. I think this is because Suppes is concerned with structure of the application of given obligations in the decision theoretic framework, rather than with the internal structure of obligation itself.

This is not an accident of Professor Suppes' approach. It is built into the Decision-Theoretic framework. Obligations are defined on *sets* of states of nature. Thus given any two obligations $f \mid A$ and $g \mid B$ (f restricted to A, and g restricted to B), we can combine them to get new obligations: e.g. $f \mid A - B \cup g \mid B$. These also belong to $\mathcal{O}$, the set of obligatory acts. In virtue of the consistency requirement we may define a Universal Obligation $o \mid X$, where X is the set of states of nature, and for all $x \in X$, $o(x) = f(x)$ if there is an A such that $x \in A$ and there is an f such that $f \mid A \in \mathcal{O}$, and $f \mid A = f(a)$ otherwise. $o \mid X$ is the universal obligatory act in the sense that every other obligation in $\mathcal{O}$ is a restriction of o. We can do all this without supposing that $o \mid X \in \mathcal{O}$ – it is still a Universal Obligation in the sense that it is all we have to know (all we ought to know!) in order to know what our obligations are: they are all restrictions of o to sets of states of nature. So it is hardly surprising that this approach doesn't throw much light on the complex inter-articulation of sometimes conflicting duties that constitutes our *real* decision problem. In some respects life is very simple in the world of Calvin!

This brings me to a point that troubled me off and on while I was reading Professor Suppes' paper. Professor Suppes admits there is a Calvinistic strain in his treatment of obligation. He says that his super-Calvinistic Theorem 16 (If Op and Pq then Oq: if there is any act which is obligatory, then every act that is permitted is obligatory) only "applies when we are given a state of affairs A and there is an obligatory act α in circumstances A. In this environment any permissible act β must be identical with α, and hence obligatory." But observe that A may be all of X and α may be the universal obligation o; or more realistically when we observe that A may be taken as the union of the domains of the partial functions in $\mathcal{O}$; or (what is the same as this last) when the event A is the set of all of those states of nature which belong to the domain of any partial obligation function whatever – this doesn't help much. But what strange bedfellows Calvinism and Decision Theory make! The Decision Theory provides us with a reasonable and workable program for acting under uncertainty. The theory of obligation tells us flatly that under every state of nature in A – which may be quite large – $o \mid A$ is the act that must be performed. Why? because it is the act that must be performed in event A. So far as the theory is concerned this is completely arbitrary and categorical. This is hardly in the spirit of Bayesian Decision Theory. This is particularly the case on Professor Suppes' view of Decision Theory. Professor Suppes, more than anyone else, has been concerned to keep the ingredients of Decision Theory tied down to empirical realities. He is a subjectivist with regard to probability for this reason. In ordinary Decision Theory, then, we have both utilities and probabilities that can be checked against experience: you can get at both utilities and probabilities by looking – essentially – at a person's preference ranking among acts. Of course, one can't really *look* at the whole preference ranking – but one can sample here and there and get a pretty good idea of what the preference ranking looks like.

Now let us perform the marriage of Suppes' theory of obligation and decision theory. We are to put some partial acts into the picture. Suppes gives a hint as to how these acts get into the picture: "... whenever there are any acts that are obligatory and the events on which they are conditioned have non-null probability, then obligation dominates all other considerations of utility." How can we tell when this is the case – i.e., when the subject of the theory regards an act as obligatory? There seems to be no way at all – at least as long as we accept the conventional assumptions that Utility is unbounded and that degrees of belief arbitrarily close to zero and one are admitted. There seems to be no behavioristic tie between Decision and Theory and the theory of obligation.

Therefore if we wish to join the two together, we must do so by a kind of marriage in name only. And as is clear from Professor Suppes' paper, after we have done so, it is easy enough to put them asunder again. Let D_o be the union of the domains of definition of the elements of $\mathcal{O}$. If the event D_o has zero probability, we can forget about it, and do our usual decision theoretic thing on $X - D_o$. If the event D_o has finite probability only one narrow path lies before us: do o. We are left with the same question that bothered Suppes when he reads Prichard: "What can be said about duty in the face of partial ignorance?"

Thus it doesn't seem to me that we have so much introduced a new dimension to decision theory as isolated – somewhat artificially – a special set of applications. We have certainly not got any closer to unravelling the tangled skein of interlocking obligations that we face as people. We couldn't have foreseen this, perhaps, without Suppes' elegant attempt to tie decision and obligation together. It doesn't seem to me, though, that the attempt succeeds in any other way. At least not yet. If Professor Suppes could answer my two main questions: viz., How do we tie obligation to Decision Theory in a fundamental way – e.g., behavioristically – and How do we avoid or reconcile ourselves to Calvinism – then I would feel better.

The University of Rochester

PATRICK SUPPES

REPLY TO PROFESSOR KYBURG

I am afraid that Professor Kyburg and I are being far too amiable in our exchange. Under the ordinary rules of the game we have an obligation to disagree! We mainly agree on the criticisms of my approach. One way out is through the concept of expected seriousness mentioned in several places in my paper, but not made part of the formal theory. There is a difficulty about this concept as well, but let me first sketch how the theory might be developed.

In almost complete analogy with the theory of expected utility we develop a theory of expected seriousness of obligation, or, what I shall term briefly, a theory of expected obligation. Following the line of attack on conditional expected utility in Krantz *et al.* (1971, Chapter 8), we can strengthen the axioms given earlier to prove the following expected obligation theorem:

There is a probability measure P on the subsets of X and a real-valued function e (e for expected obligation) such that for all f_A and g_B in $\mathcal{O}$

(i) $f_A \geq g_B$ if and only if $e(f_A) \geq e(g_B)$,
(ii) If $A \cap B = \emptyset$ then
$e(f_A \cup g_B) = e(f_A) P(A \mid A \cup B) + e(g_B) P(B \mid A \cup B)$.

I shall not discuss the axioms that are needed to prove such a theorem. In one form or another, however, the relative seriousness of obligations must be compared and weighted with the probability of relevant circumstances occurring. No doubt in a qualitative way such judgments are involved in resolving the kind of conflict mentioned by Kyburg.

I do not mean to suggest, by the way, that the introduction of a theory of expected obligation really provides all the machinery for the actual resolution of conflicts in obligations. Only a general framework is provided by the ideas I have suggested. Substantive principles need to be formulated inside the framework before a theory of obligation of any real substance has been given.

There is a general difficulty, however, about the framework itself that I want to examine. It is this. Suppose a young man is invited by a young woman to spend a weekend skiing and doing other enjoyable things, the total anticipated utility or pleasure being considerable. On the other hand, his mother is mildly ill with a still undiagnosed ailment. Suppose the event A of her dying during the course of the weekend is less than 10^{-6} in his subjective estimate. He accepts without question the obligation of being at her bedside when she is dying if it is humanly possible for him to be there. Let f be the act of staying home, and g the act of going away for the weekend with the young woman. Let, for simplicity,

$$f_{\neg A}, g_{\neg A} \notin \mathcal{O},$$

i.e., if his mother is not dying, event $\neg A$, neither f nor g is obligatory. Let now u be expected utility and let

$$u(g_{\neg A}) \gg u(f_{\neg A}),$$

i.e., the utility or pleasure of g given $\neg A$ is very much greater than the pleasure of f given $\neg A$.

The theory of expected obligation is not adequate to the situation, nor to any other situation in which expected utility and expected obligation must be compared, or, perhaps it is better to say, the probability of any obligatory act needing to be performed is low enough that utility considerations become pertinent.

From a formal standpoint it would be easy enough to resolve the whole issue by just considering expected utility, but many of us feel genuine decisions are made on a two-track model of obligation and utility, and one should not be totally absorbed by the other.

One solution is to introduce the concept of a threshold of obligation. In the present example if the young man finds $e(f_{\neg A})$ below threshold, or qualitatively, just $f_{\neg A}$ itself, then obligation can be ignored and the choice made on grounds of utility or pleasure. It is important to note that an obligation can fall below threshold for two interlocking reasons: it is not a very serious obligation or the chances of the circumstances occurring that require its observance are low. Of course, moderate obligations with moderate chances of being applicable can fall below threshold because of the interaction of the two sorts of 'moderateness'.

From a formal standpoint the idea of such a threshold can be worked out in a number of ways, which I shall not pursue here. To answer Professor Kyburg's last question it is my present feeling that the three concepts of expected obligation, expected utility and threshold of obligation can form the framework of a more realistic, necessarily more complex and also a behaviorally testable theory of obligation. But to express this feeling is far short of offering the theory itself in satisfactory form.

BIBLIOGRAPHY

Krantz, D. H., Luce, R. D., Suppes, P., and Tversky, A., *Foundations of Measurement*, Vol. 1, Academic Press, New York, 1971.

R. B. BRAITHWAITE

BEHIND DECISION AND GAMES THEORY:
ACTING WITH A CO-AGENT VERSUS
ACTING ALONG WITH NATURE

The work to be outlined here has arisen out of my attempts, over the last decade, to say something definite, even if only schematic, about the connexion between belief and action. Most philosophers agree that the relation between having a belief and, in appropriate circumstances, acting on the belief cannot be an accidental or even merely a causal connexion. The belief in its context supplies a good and sometimes an adequate reason for the action; and this is a logical and not an empirical fact. But the relation of belief to action is complicated by the fact that it cannot be considered independently of the believer's system of preferences: a belief that a river is deep is a good reason for a man who cannot swim and who is intent on suicide to jump into it, and a good reason for him not to jump into it if he has no such intent. What is needed for the analysis of the relations between a man's beliefs, his preferences and his actions is to find some way of distinguishing the part played by his beliefs from the part played by his preferences in guiding his choices of action. If it were possible to make this discrimination, it would then be profitable to consider what about a man's beliefs and preferences can be deduced from his actions; and this, for me, would be to complete the pragmatists' programme of determining what a man is (in the way of believing and preferring) by finding out what he does.

After going up several blind alleys in my search for a discriminating principle, six years ago the idea came to me of examining the whole problem in the light of the fundamental ideas of Decision Theory and Theory of Games. Both these theories are concerned with propounding and discussing principles for rational choice of action when the choice is 'under uncertainty'; i.e. when the outcome of the agent's action depends jointly upon what action he chooses and upon another factor about whose occurrence the agent is uncertain. Decision and Games Theorists represent the choice-situation by a rectangular array with lm cells arranged in l rows and m columns, the l rows representing the l possible actions of the agent, the m columns representing the m possible alternative

Leach et al. (eds.), Science, Decision and Value, 22–55. All Rights Reserved.
Copyright © 1972 by D. Reidel Publishing Company, Dordrecht-Holland.

factors. In Decision Theory these alternative factors are alternative propositions, when the agent is uncertain as to which of them is true; in Theory of Two-Person Games (the only part of Games Theory with which I shall be concerned) the alternative factors are alternative actions of another agent. In Decision Theory, the entities in the cells of the array are single entries representing in each cell the relevant part of the preference system of the one agent (whom I shall call Luke throughout). In (Two-Person) Games Theory they are double entries representing the relevant parts of the preference systems of both Luke and the other agent (whom I shall throughout call Matthew). The features common to both theories are that the preference systems are taken as known data, and that the choice-situations (for Luke in Decision Theory, for both Luke and Matthew in Games Theory) are those of choice under uncertainty. Because of these common features Decision Theorists frequently refer to the choice-situations with which they are concerned as 'games with nature' (sometimes, alas, as 'games against Nature'). I shall follow them in using 'game' to include 'game with nature' for the sole but sufficient reason that the spoken word 'game' is a monosyllable and the written word a four-letter word, neither of which is the case for the expression 'choice-situation'.

To start with absence of certainty as a datum is of course perfectly proper for a Decision or Games Theorist, since it was to deal with the very difficult problem of how to choose rationally under uncertainty that the two theories were created. But their rectangular-array representation devised for considering their own problem suggested to me that it might help in considering my belief-preference-action problem in the following way. Can one not go, as it were, behind Luke's choice-situation-under-uncertainty corresponding to a game represented by an array with m columns to Luke's proto- or *Ur*-choice-situation which is in itself a situation neither of certainty nor of uncertainty (call it a situation of *nescience*) and which will correspond not to the *total game* whose array has m columns but to the *set of games* each game of which is represented by an array with one or more of the m columns, i.e. in a pretty obvious sense, the complete set of *subgames* of the total game, the subgames (when we are considering Luke's situation) all having the same l rows in their arrays as the total game? If one can do this, the process of Luke's choice of action can be divided into two distinct logical steps. First a game will be selected out of the complete set of games which is Luke's *Ur*-choice-situation; then Luke will choose an action (represented by his choosing a row) in the game thus selected. And perhaps it will be the case that the first step can be regarded as the contribution made to the whole choice-process by Luke's certainty of one of the alternatives or his uncertainty between two or more of them (for convenience I shall call both his certainty and his uncertainty a *credal state* of Luke's), the contribution made to the whole process by Luke's preference system being confined to the second step, where it will function in the way discussed by Decision and Games Theorists. Then the way in which Luke's choice of action is guided by his credal state will be distinguished from the way in which it is guided by his preference system by means of the distinction between the two steps. Luke's credal set will determine the game selected from the complete set of games of his *Ur*-choice-situation; Luke's preference system will then guide his choice of action in the selected game. Section I of this paper will outline a schematized theory of the functioning of the first step in Luke's games with nature. Section II will apply a modified theory to the first step in Luke's games with Matthew, with a result which surprised me very much. In both these parts we shall be concerned only with Luke's credal states and principles for choice of action: Matthew will only come in by way of Luke's credal states about him. In Section III an observer of Luke's behaviour will be introduced – a sociologist or behavioural psychologist or perhaps an animal-behaviour man called Karl: he will have to make hypotheses about Luke's own system of preferences, which will not have come into the argument before.

I

The simplest case of a game of Luke's with nature is one in which there are only two propositions concerned, which are p, not-p. If M_1, M_2 are the columns corresponding to p, not-p respectively, the total game may be regarded as the class of these two columns $[M_1, M_2]$ and the three subgames as the one-membered classes $[M_1]$, $[M_2]$ together with $[M_1, M_2]$ itself. (The null-class comprising no column will not be regarded as a subgame.) According to my theory as I first thought of it, Luke's certainty of p will determine that he plays (i.e. makes his choice of row) in game [M₁], his certainty of not-*p* that he plays in game [M₂], his uncertainty between *p* and not-*p* that he plays in the total game [M₁, M₂].

So far, so good for my theory in its original form. But it will not work satisfactorily if the total game has more than two columns. When the total game has three columns M_1, M_2, M_3 corresponding respectively to *p*, *q*, neither *p* nor *q*, with *p*, *q* logically disjoint, if Luke's certainty of *p* yields game [M₁] for him to play in, his certainty of *q* yielding game [M₂] and his uncertainty between *p* and *q* yielding game [M₁, M₂], what is the game for him to play in if he is certain of *p* ∨ *q* without being certain either of *p* or of *q*? And what is the game for him to play in if he is uncertain between the truth and the falsity of *p* ∨ *q*? All the seven subgames are already required for the games yielded by Luke's certainties of the three propositions separately, by his uncertainties between them taken in pairs, and by his uncertainty between all three of them.

The only way to answer these questions is to modify the theory from its original simple-minded form, and to regard the function of Luke's credal state as being not that of selecting a *subgame* out of Luke's *total* game, but that of selecting a *subset* of subgames out of the *complete set* of subgames of this total game. With this modification the theory works most satisfactorily. The *credal sets* (as I shall call them) determined by Luke's certainty of any one of the three propositions or by his uncertainty between any two or between all three of them can be taken to be the seven one-membered sets included in the complete set. Then Luke's certainty, for example, of *p* will yield the set [[M]] instead of the game [M₁]; his uncertainty between *p* and *q* will yield [[M₁, M₂]] instead of [M₁, M₂]; and in all seven cases Luke's credal state will determine a unique game for him to play in, since in each case the credal set will have only one game belonging to it. But since there are $2^7 = 128$ subsets of the complete set, one of which is the null-set and seven of which are the one-membered sets, there are 120 many-membered sets from which to provide credal sets for the more complicated credal states of Luke. This, as we shall see, is more than enough.

However the introduction into my theory of credal sets which comprise more than one game raises a nasty problem which I shall sidestep rather than solve. For if the function of Luke's credal state, in guiding his choice of action, is to select a subset out of his complete set of games, and this subset (unless it is one of the one-membered sets) will comprise several games, what is it that will select out of this subset the unique game in which Luke is to play? Is not a third step required in my analysis of Luke's whole choosing process, interposed between the two steps I have emphasized?

No extra step is called for if Luke, besides holding the creed of his being certain of $p \vee q$ (I shall use the monosyllabic word 'creed' for a proposition asserting that a specific person has, or is in, a specific credal state) also holds the stronger creed of being certain of p. In this case his weaker creed will restrict the game in which he is to play from being merely one of the complete set (corresponding to a state of nescience) to being one of the set $[[M_1], [M_2], [M_1, M_2]]$ (this will be shown later). His stronger creed will further restrict his game to being comprised in the set $[[M_1]]$ i.e. to being $[M_1]$. So being presented with a many-membered credal set may stimulate Luke to examine whether he does not hold a stronger creed which will settle for him a unique game.

And it will occasionally be the case that Luke's principle of rational choice is such as to direct him to choose the same action (the same row) whichever of the games in his credal set he is to play in. Here a selection of a unique game from the credal set will be unnecessary, and Luke can (as it were) play *to* the credal set instead of playing *in* a unique game belonging to the credal set.

But these considerations do not really solve the nasty problem for me produced by the plurality of games in my credal sets. However I shall sidestep the problem by saying that it is peripheral to the main purpose of this paper, which is not to analyze the psychology of deliberation in choice of action but to give such an account of the logical relations between credal states, preference systems and choices of action as to enable creeds and propositions about preferences to be subject to some sort of behavioural test. The behavioural test can only be, even under the most favourable circumstances, that of refuting the hypothesis that Luke holds such-and-such a creed yielding such-and-such a credal set; and in this testing the 'micro-structure' of Luke's process of choice will be irrelevant. In commenting upon the paper which I read at the American Philosophical Association Western Division meeting at St. Louis last year, Isaac Levi said that my notion of credal set would be unobjectionable if the credal set were thought of as "the set of games which we, as observers of Luke's behaviour, have failed to rule out as *the* game which Luke is facing". With minor qualifications I gladly accept this way of regarding my theory. What I am concerned with is to specify credal sets in terms of creeds in such a way that a testing of the former can also be a testing of the latter: for this testing what happens, as it were, *within* a credal set is irrelevant.

The implicit reference to a possible observer will explain the assumption about Luke's reasonableness that I shall make, Decision Theorists have suggested and defended different sophisticated principles for rational choice under uncertainty which are mutually inconsistent and each of which has paradoxical consequences: Patrick Suppes has compared the situation in Decision Theory to that confronting Set Theory at the turn of the century. It would be ridiculous for me to produce a theory which would require an observer to assume that Luke was using as his principle of rational choice the Maximin Principle, for instance, or that of Minimax Regret or of the maximizing of expected utilities. However there is one of the Decision and Games Theorists' principles of choice which is simple and, in most contexts, inexceptionable, namely the Domination Principle. (Richard Jeffrey has given examples when the propositions with which Luke's creeds are concerned are probabilistically dependent upon the action he chooses where use of the Domination Principle would lead to unpalatable results: I exclude such propositions from my scheme. John Harsanyi has published a general method of solution for non-zero-sum non-cooperative games which in some cases contravenes the Domination Principle: I am sceptical of the reasonableness of his solution.) The Domination Principle is to the effect that a rational man, in a situation of choice under uncertainty, will not choose one action rather than another if he *may lose* but *cannot gain* by doing the first action rather than the second. This principle is too weak to be of much assistance in Decision and Games Theory, but I found, somewhat to my surprise, that it was strong enough for my purpose. So the only assumption to be made about Luke's reasonableness (in choosing in the game in which he is playing) is that he will not choose an action which is dominated by another of his possible actions.

The fact that the Domination Principle is the only principle of choice required for my theory had two important consequences for my formal development of it. First, use of the Domination Principle requires only that the entries in each column in Luke's array should have a preference structure, i.e. admit of comparisons of preference with one another: it does not require any cross-column preference comparisons, still less any assignment of numerical 'utilities' to outcomes represented by the entries. Moreover the entries need not be taken as referring to 'outcomes' in the sense either of limited consequences or of possible future states of the universe (and each of these senses raises serious difficulties). All that is necessary is that the entries in the column M_s corresponding to (say) the proposition m_s should indicate for each pair of actions e.g. L_1, L_2, one or other of three alternatives, taken to be exclusive and exhaustive: (1) that Luke would prefer to do L_1 rather than to do L_2 were m_s to be true, (2) that Luke would prefer to do L_2 rather than L_1 were m_s to be true, (3) that Luke would prefer neither action to the other were m_s to be true; with the added qualification in all three cases that his preference or lack of preference would be unaffected if a conjunction of m_s with any proposition logically consistent with it and satisfying two minor conditions (to exclude Jeffrey's cases and iterated credal modalities) were substituted for m_s. All that will be presupposed in my argument is that there is a single preference structure $\mathscr{R}_s$ in each column M_s separately. These single preference structures need not in general be transitive, thus giving rise to a simple ordering, though in particular games transitivity for some parts of the structure may be essential in order to ensure that in every game there is at least one undominated action.

The second important consequence of my use of only the Domination Principle is that, since it requires no assignment of numerical utilities, I shall assign no numerical measurement within my single preference structures and shall employ only the very weak logic of preference outlined. And this makes it pointless for me to discriminate degrees of Luke's belief within the range of his uncertainty: there would be no object for me in bringing into my theory Luke's subjective probabilities (his betting quotients) when I have no numerical utilities to multiply them with. Of course it is essential that everything I say about Luke with my weak logic should be consistent with what would come out of a stronger logic, which Luke may very well have. For example, if Luke assigns numerical probabilities to measure his degrees of uncertainty and numerical utilities to the entries in each column, he will be able to use, if he wishes, the maximizing-of-expected-utilities principle as a principle of choice. But my use of only the Domination Principle will not be inconsistent with his use of the stronger principle, from which indeed the Domination Principle logically follows. No action forbidden by the Domination Principle will fail to be forbidden by his principle, and it is the inadmissibility of actions according to the Domination Principle that will enter my argument. In fact all my results can be interpreted quite elegantly in a logic which has room for numerical subjective probabilities, certainty of p corresponding to the probability of p being 1, certainty of not-p to the probability of p being 0, uncertainty between p and not-p corresponding to the probability of p lying in the open interval $(0, 1)$.

Before sketching the formal development of my theory, I must explain the types of creed whose ability to guide action by determining credal sets will be explained. I shall take creeds to be of one or other of the following forms:

> Luke is certain of p – abbreviated into $C_L(p)$
> Luke is uncertain between p and q, where p, q are logically disjoint propositions – abbreviated into $U_L(p, q)$
> A conjunction of creeds of one or other or both of these forms.

This specification of creed deliberately excludes absence of certainty, as also absence of uncertainty, from being a creed. My reason for this is primarily in order that the behavioural consequences which my theory deduces (of course when many assumptions are made) from Luke's holding any specific creed should be *necessary* but never *sufficient* conditions for his holding the specific creed. And this is essential for me to escape the accusation that I am trying to *reduce* belief to behaviour. It is no part of my thesis that there is nothing to belief except its function in guiding action: so far as my theory is concerned a belief or any other credal state may contain private mental states or tacit speech-acts or physical events in a brain or other part of the body. The essential part of my thesis is that holding a belief (a 'genuine' belief, a 'sincere' belief) guides action in appropriate circumstances. If absence of belief were also to guide action, it would be possible, given some assumptions, to deduce from behaviour inconsistent with such action that the man positively held a belief. My theory permits of no such inferences: all it permits are inferences to negatives of creeds.

This refusal of mine to count negatives of creeds as creeds requires however that I should regard uncertainty between two disjoint propositions p, q, as a positive credal state sui generis, and not as an absence of certainty of p combined with an absence of certainty of q. Here I can claim the support of common sense, or (better to say) of introspective or armchair psychology. The positive and agonizing character of uncertainty about the existence of God, about survival after death, about human destiny, has been eloquently described by William James and many novelists: it is difficult to see how a mere absence of a state of certainty could generate so much *Angst*. So I feel justified in taking being uncertain about p as meaning being uncertain between p and not-p, which is quite different from not being certain of p and not being certain of not-p. Alternations of creeds as themselves creeds I reject for somewhat similar reasons.

Uncertainty has been taken to be always between two disjoint propositions, for it is only when p and q are disjoint that the sentence 'Luke is uncertain between p and q' has an unambiguous sense. Even when three propositions p, q, r are pairwise disjoint, the sentence 'Luke is uncertain between p, q, r' may be ambiguous. If it is taken to mean, using my abbreviations, $U_L(p, q \vee r) \wedge U_L(q, p \vee r) \wedge U_L(r, p \vee q)$, it is a conjunction of creeds and thus a creed without having to be allowed for separately.

In developing my theory I found that it was not necessary to posit any postulates of a modal logic relating certainties and uncertainties among themselves, not even the uncontroversial equivalence of $C_L(p) \wedge C_L(q)$ to $C_L(p \wedge q)$, if I was prepared not to use iterated certainties and uncertainties, such as $C_L(C_L(p))$, $C_L(U_L(p, q))$ in my logical system. This seemed a cheap price to pay for avoiding the complexities and controversies of modal logic, and I have paid it. The propositions which Luke's creeds are about will therefore include no creed held by Luke. They may include creeds held by Matthew, since $C_L(C_M(p))$ is not an iterated modality.

My logical system requires six postulates to relate credal sets to one another. Small Greek letters α, β, γ, μ, ν, etc. will be used to stand for games (each of which is the class of its columns), capital Greek letters

A, B, Γ, etc. to stand for sets of games. The credal set of a creed q held by Luke will be written as $\Lambda^c q$.

POSTULATE 1. If p_1 is logically equivalent to p_2, $\Lambda^c C_L(p_1) = \Lambda^c C_L(p_2)$; and if p_3 is logically disjoint to p_1 (and therefore also to p_2), $\Lambda^c U_L(p_1, p_3) = \Lambda^c U_L(p_2, p_3)$

POSTULATE 2. For any creeds q_1, q_2, $\Lambda^c(q_1 \wedge q_2) = \Lambda^c q_1 \cap \Lambda^c q_2$

POSTULATE 3. $\Lambda^c C_L(p_1 \wedge p_2) = \Lambda^c C_L(p_1) \cap \Lambda^c C_L(p_2)$

For the next two postulates I found it most convenient to define two novel logical operations upon sets. (The word *set* will always only be used for classes of the second order, which will usually be classes of games, themselves classes of the first order).

The *overlap* of A and B, written as $A \bigcirc B$, is defined as the set each of whose members is the union of one member of A and one member of B.

Since all the sets with which we shall be concerned will comprise a finite number of classes as members, the Axiom of Choice is not involved; and for our finite cases the definition can be given formally as:

When $A = [\alpha_1, \alpha_2, \ldots \alpha_l] = U_i[\alpha_i]$, $B = [\beta_1, \beta_2 \ldots \beta_n] = U_j[\beta_j]$
$$A \bigcirc B = U_i(U_j[\alpha_i \cup \beta_j])$$

Simple cases:

$$[\alpha] \bigcirc [\beta] = [\alpha \cup \beta], [\alpha] \bigcirc [\beta_1, \beta_2] = [\alpha \cup \beta_1, \alpha \cup \beta_2]$$

The *conflation* of A and B, written as $A \oplus B$, is defined as $A \oplus B = A \cup B \cup (A \bigcirc B)$

Simple case:

$$[\alpha] \oplus [\beta] = [\alpha, \beta, \alpha \cup \beta]$$

Now we can state

POSTULATE 4. If p_1, p_2 are logically disjoint propositions,

$$\Lambda^c U_L(p_1, p_2) = \Lambda^c C_L(p_1) \bigcirc \Lambda^c C_L(p_2)$$

POSTULATE 5. For any propositions p_1, p_2,

$$\Lambda^c C_L(p_1 \vee p_2) = \Lambda^c C_L(p_1) \oplus \Lambda^c C_L(p_2)$$

Finally, for one stage in the argument we require

POSTULATE 6. For any proposition p, $\Lambda^c C_L(p)$ is the null-set only if p is logically impossible. This postulate secures that there should always be at least one game for Luke to play in.

These postulates enable all the credal sets concerned in Luke's games with nature or with a person to be constructed out of the fundamental credal sets for the games in question.

The fundamental credal sets for games with nature are Luke's certainties of m propositions $m_1, m_2, \ldots m_m$, which are pairwise disjoint and such that $m_1 \vee m_2 \vee \ldots m_m$ is logically necessary, occurring in the *basic choice-situation*, the l-rowed game with m columns $M_1, M_2, \ldots M_m$ corresponding respectively to the m propositions. For the choice-situation to be *basic*, it is necessary that every column M_s should have a single preference structure $\mathscr{R}_s$ given by Luke's preference or lack of preference between every pair of his actions were m_s to be true, independently of whether any other proposition (with minor exceptions) were also true. The *basic decomposition* $[m_1, m_2, \ldots m_m]$ is supposed to be sufficiently fine for every column to have a single preference structure; this must be given as a postulate in a proper formal treatment.

Taking $[M_1] = \mu_1, [M_2] = \mu_2, \ldots [M_m] = \mu_m$, the basic choice-situation σ is the union of these *monadic games*:

$$\sigma = [M_1, M_2, \ldots M_m] = [M_1] \cup \ldots [M_m] = \mu_1 \cup \mu_2 \cup \ldots \mu_m.$$

The monadic games are minimals in the system, every polyadic game being expressible uniquely as a union of distinct monadic games. The complete set of the subgames of σ will be written as $\Theta_{12\ldots m}$: it is the power set of σ with the null-class omitted.

The fundamental credal sets for games with nature are given by the *Application Postulate*:

For every s, $s = 1, 2, \ldots m$, $\Lambda^c C_L(m_s) = [\mu_s]$.

Thus the credal set of $C_L(m_s)$ is the set whose sole member is the monadic game μ_s, whose sole member is the column M_s.

Every credal set for games with nature is obtained by performing a sequence of operations of conflation, overlap and interesection upon some or all of these fundamental credal sets. The conflation and intersection operations can be repeated, but the overlap can only be done once and can only be followed by an intersection. Conflation forms new credal sets by Postulate 5, overlap by Postulate 4, intersection by Postulates 2 and 3.

To work this out requires using the formal properties of the overlap and conflation operations. Each of these operations is associative, commutative, distributive over union, and monotonic.

For every A, A ⊆ A ○ A ⊆ A • A; for every A, B, A ∩ B ⊆ A○B ⊆ A•B ⊆ A∪B.

But in general neither operation is idempotent. However there are many sets A for which $A = A \bigcirc A$ and $A = A \oplus A$, these identities being logically equivalent: such sets will be called *perfect*. Every one-membered set is perfect; and if A, B are perfect, so also are $A \bigcirc B$, $A \oplus B$, $A \cap B$. Thus every credal set for a game with nature is perfect, since it is constructible by these three operations from the one-membered fundamental credal sets.

But a stronger theorem holds, namely that the credal set of every certainty is perfect, and hence *every credal set is perfect*, since it is constructible by the three operations from credal sets of certainties. (*Proof.* $\Lambda^c C_L(p) = \Lambda^c C_L(p \vee p)$, by Postulate 1. But $\Lambda^c C_L(p \vee p) = \Lambda^c C(p) \oplus \Lambda^c C_L(p)$, by Postulate 5.) This raises the question as to whether every possible credal set, that is, every perfect set, can occur in Luke's games with nature, and we will consider this by calculating the credal sets that can so arise when Luke's choice-situation has three columns so that his total game σ is $\mu_1 \cup \mu_2 \cup \mu_3$. Generalization to any number of columns greater than 3 raises only computational difficulties.

Certainties or uncertainties, or conjunctions of these, about some or all of the propositions m_1, m_2, m_3 yield credal sets as shown below. Here, and in what follows, the one-membered sets included in the complete set Θ_{123} are denoted by single numerals: $1 = [\mu_1]$, $2 = [\mu_2]$, $3 = [\mu_3]$, $4 = [\mu_1 \cup \mu_2]$, $5 = [\mu_1 \cup \mu_3]$, $6 = [\mu_2 \cup \mu_3]$, $7 = [\mu_1 \cup \mu_2 \cup \mu_3]$, concatenations of these numerals denoting the other sets included in Θ_{123}, e.g. $124 = [\mu_1, \mu_2, \mu_1 \cup \mu_2]$; $1234567 = \Theta_{123}$.

$C_L(m_1), C_L(m_2), C_L(m_3)$ → 1, 2, 3
(the fundamental credal sets)
$U_L(m_1, m_2), U_L(m_1, m_3), U_L(m_2, m_3)$ → 4, 5, 6
$C_L(m_1 \vee m_2), C_L(m_1 \vee m_3), C_L(m_2 \vee m_3)$ → 124, 135, 236
$C_L(m_1 \vee m_2 \vee m_3)$ → 1234567
(this imposes no restriction on Θ_{123})

$$U_L(m_1, m_2 \vee m_3), U_L(m_2, m_1 \vee m_3), U_L(m_3, m_1 \vee m_2)$$
$$\rightarrow 457, 467, 567$$

$$\left.\begin{array}{l} U_L(m_1, m_2 \vee m_3) \wedge U_L(m_2, m_1 \vee m_3) \\ \text{and two similars} \end{array}\right\} \rightarrow 47, 57, 67$$

$$U_L(m_1, m_2 \vee m_3) \wedge U_L(m_2, m_1 \vee m_3) \wedge U_L(m_3, m_1 \vee m_2)$$
$$\rightarrow 7$$

These seventeen credal sets are all that arise if the propositions concerned in Luke's creed (call these the *credal propositions*) are restricted to m_1, m_2, m_3. A further sixteen credal sets arise when the credal propositions are some or all of the propositions in another decomposition $[p_1, p_2, \ldots p_n]$ where some or all of these propositions are distinct from but logically related to some or all of the m_s's. These credal sets are 14, 15, 24, 26, 35, 36, 1457, 2467, 3567, 4567, 14567, 24567, 34567, 124567, 134567, 234567. The family $F_N(3)$ of the credal sets included in Θ_{123} which are obtainable in games with nature thus has 33 members. It does not include all the sets which are perfect, for this family $P(3)$ has 60 members, not counting the null-set.

The family $F_N(3)$ can be constructed out of the family $[1, 2, 3]$ of the fundamental credal sets (included in Θ_{123}) for games with nature by a sequence of two closure operations. First construct the conflation closure of $[1, 2, 3]$, which corresponds to constructing $\Lambda^c C_L(x \vee y)$ as $\Lambda^c C_L(x) \oplus \Lambda^c C_L(y)$ according to Postulate 5. This conflation closure is $[1, 2, 3, 124, 135, 236, 1234567]$. Second, construct the overlap closure of this family, which corresponds to constructing $\Lambda^c U_L(x, y)$ as $\Lambda^c C_L(x) \bigcirc \Lambda^c C_L(y)$ according to Postulate 4. This overlap closure is the family $F_N(3)$ comprising the 33 sets we have specified. (Why an intersection closure is not included in the sequence of two closure operations is that an intersection closure either at the beginning or in the middle or at the end of the sequence would add only the null-set.)

The family $F_N(3)$ has an interesting lattice-theoretic characterization (which I have proved in general though I shall explain it only for our case of $m = 3$). The 33 sets belonging to $F_N(3)$ can be partitioned into seven subfamilies $D_1, D_2, \ldots D_7$ in the following way: D_n, for $n = 1, 2, \ldots \ldots 7$, is to be the family of sets each of which includes the set denoted by n and is included in the complete set of the subgames of the game which is the unique member of the set n. (For examples, $D_1 = [1]$; $D_4 =$

= [4, 14, 24, 124].) Each of the 33 sets of $F_N(3)$ falls into one and only one of these subfamilies. It can be proved that, if A, B belong to the same subfamily D_n, $A \bigcirc B = A \cap B$ and belongs to D_n, $A \oplus B = A \cap B$ and belongs to D_n, from which it follows that each subfamily is a free distributive lattice with union and intersection as the lattice operations. The Hasse diagrams of the seven free distributive lattices are shown on the figure (see p. 36), with the 33 sets specified as vertices (D_1, D_2, D_3 have each only one set as member). The sets marked with an X are the 'generators' of the free distributive lattice to which they belong: each lattice is constructible by repeated union and/or intersection of its generators, and is closed with respect to each of these operations.

II

We will now consider Luke's choice situation when the other factor determining the outcome of his choice of action is the choice of action of another person Matthew who is also in a choice-situation – the subject-matter of Two-Person Games Theory. This theory was initiated in the 1920's by Borel and von Neumann (independently) by making the reasons which Matthew would have to choose one action rather than another the basis for the calculations which Luke would make in deciding how to act, and *vice versa*. This way of looking at the matter is the *differentia* which, within the *genus* of Luke's games under uncertainty, sharply distinguishes the *species* of his games with a person from the *species* of his games with nature. In playing a game with nature Luke's fundamental credal sets $[\mu_1], [\mu_2], \ldots [\mu_m]$ from which all his other credal sets are derived correspond each to a certainty that a particular one of the propositions belonging to the basic decomposition $[m_1, m_2, \ldots m_m]$ will be true. If in playing a game with Matthew Luke were to calculate on the basis of these being the fundamental credal sets, he would be presupposing the *possibility* of his being certain, for example, of m_1, which is the proposition that Matthew will perform the action M_1 i.e. choose the column M_1. But if Luke *presupposes* this possibility in his calculations, he will be treating his game with Matthew as being a game with nature; and if he treats it thus, it *will be* a game with nature in which 'Matthew' is his nickname for an automaton. A game is what the player takes it to be. So $[\mu_1], [\mu_2], \ldots [\mu_m]$ cannot be the fundamental

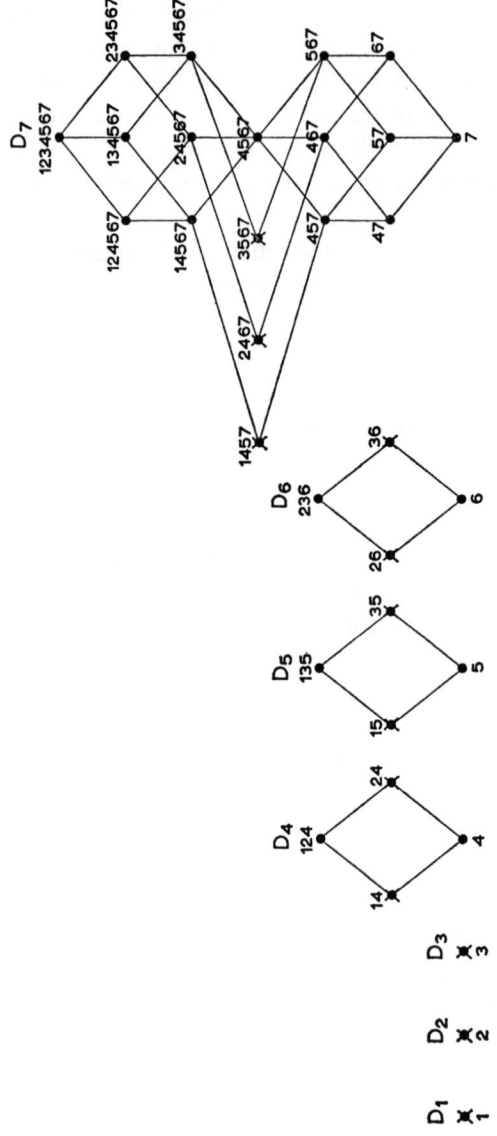

credal sets in Luke's game if he is genuinely playing a game with Matthew.

The fact that in Luke's game with Matthew $[\mu_1]$, for example, cannot be a fundamental credal set by no means prevents it from being a credal set. I shall be explaining two rules to be used for determining what are to be fundamental credal sets in a game with a person, and in both of these the set $[\mu_1]$ will appear as the credal set of a certainty of a proposition in the relevant decomposition. It will therefore be called a quasi-fundamental credal set. But it will not be a fundamental credal set, and it will appear as the intersection of fundamental credal sets. If Luke's certainties about Matthew's preference system etc. yield for Luke $[\mu_1]$ as the credal set, Luke's certainties about Matthew's preference system etc. will guide his choice of action in exactly the same way *as if* he were certain that Matthew would choose M_1. Indeed there is no objection to Luke's being certain of this and still be playing a genuine game with Matthew, provided that the explanation Luke would give of $[\mu_1]$ being the credal set guiding his choice of action were not only that he was certain that Matthew would choose M_1 but also that the basis for this certainty was a certainty of considerations about Matthew's preference system etc, which are sufficient in themselves to yield $[\mu_1]$ as his credal set.

What considerations about Matthew's preference system etc. must be credal propositions in Luke's creeds if he is to be playing a (genuine) game with Matthew? It is essential that one of these credal propositions V_M should describe all of Matthew's preference system that could be relevant to his choice, and this requires that V_M should specify the single preference structure $\mathscr{S}_r$ (for Matthew) in each column L_r of Matthew's basic choice-situation $\tau = [L_1, L_2, \ldots L_l] = \lambda_1 \cup \lambda_2 \cup \ldots \lambda_l$, where the λ_r's are Matthew's monadic games. (It will be assumed that τ is a basic choice-situation for Matthew just as σ is a basic choice-situation for Luke.) But the argument requires that Luke should hold a creed of another sort about Matthew, and what this creed should be allows some freedom of choice to the philosophical analyst. It is *prima facie* plausible to take this creed as being a certainty or uncertainty of Luke's as to Matthew's certainty or uncertainty as to what action L_r (which row L_r) Luke will choose. But, on my view, to discuss the matter in this way will presuppose that, while Luke is playing a game with Matthew, he believes that Matthew is in fact playing a game with nature and treating Luke as an automaton. This, of course, is one possibility that must be allowed for. Another possibility is that Luke believes that Matthew is playing a genuine game with Luke and is basing his calculations as to what column to choose on considerations as to what reason Luke might have for choosing one row rather than another. But this possibility will again divide into two possibilities – that Luke believes that Matthew believes that Luke is playing a game with nature, and that Luke believes that Matthew believes that Luke is playing a game with a person. And this latter possibility will again divide into two, and so on. So discussing the subject in this way will lead to an infinite regress of possibilities all of which will have to be considered. Fortunately there is another way of discussing the matter. Whether Luke believes that Matthew is playing a game with nature or believes that he is playing a game with a person, Luke will use his beliefs about Matthew to guide his choice of action *via* the intermediate step of inferring from his beliefs about Matthew beliefs about what Matthew's credal set will be, since the game δ in which Matthew will play will be restricted to belonging to his credal set. So the additional creed about Matthew which we shall take Luke to have will be one about Matthew's credal set. And then a further simplification is possible. Luke's creed, for example, of being certain that Matthew's credal set M is, for example, $[\delta_1, \delta_2]$, δ_1, δ_2 being subgames of $\tau = \lambda_1 \cup \lambda_2 \cup \ldots \lambda_p$, will yield for Luke exactly the same credal set as will the creed of being certain that Matthew will either be playing in game δ_1 or will be playing in game δ_2 i.e.

$$\begin{aligned}\Lambda^c C_L(M = [\delta_1, \delta_2]) &= \Lambda^c C_L(\text{Matthew plays in } \delta_1 \\ &\quad \vee \text{ Matthew plays in } \delta_2) \\ &= \Lambda^c C_L(\text{Matthew plays in } \delta_1) \\ &\quad \oplus \Lambda^c C_L(\text{Matthew plays in } \delta_2).\end{aligned}$$

But this is a conflation of two credal sets each of which is the credal set of a certainty of Luke's that Matthew plays in a particular game belonging to Matthew's credal set. So in our search for the fundamental credal sets in Luke's game with Matthew, we need only take into account the credal sets of his certainties that Matthew is playing in each particular game belonging to his credal set, since Luke's credal sets of certainties that Matthew is playing in *some* game of his credal set can be constructed out of these credal sets by means of the conflation operation. So the fundamental credal sets for Luke in his game with Matthew can be found by considering Luke's certainties of propositions each of which is the conjunction of a proposition V_M describing the relevant part of Matthew's preference system with a proposition saying in which game Matthew is playing.

The first rule for determining fundamental credal sets owes its plausibility to the following consideration. Matthew will be playing in some game included in $\tau = \lambda_1 \cup \lambda_2 \cup \ldots \lambda_l$ where $\lambda_r = [L_r]$, L_r being a row which Luke might choose. We are assuming that each L_r has a single preference structure $\mathscr{S}_r$ for Matthew. Now it may be the case that if Matthew plays in a particular game δ, his preference system V_M (for brevity I omit the words 'relevant part of') is such that one of his m possible actions (say M_s) is dominated by another possible action. This will happen if there is another of his possible actions (say M_t) which is such that M_s is not preferred to M_t in any single preference structure $\mathscr{S}_r$ whereas M_t is preferred to M_s in at least one of these single preference structures. Calling an action *inadmissible* if it is dominated by some other action, Rule A for determining Luke's fundamental credal sets is that, if Luke is certain of the proposition p that Matthew is playing in such a game and has such a preference system that the action M_s is *inadmissible* in that game and for that preference system, then Luke *must not allow for the possibility of Matthew's choosing M_s*, i.e. the game in which Luke plays must be one which does not include μ_s as a subgame. Applying this to the case in which $m = 3$ and Luke's total game is $\sigma = \mu_1 \cup \mu_2 \cup \mu_3$ (and I shall assume this to be the case in what follows), since $[\mu_s] \bigcirc \Theta_{123}$ is the set of games belonging to Θ_{123} which include μ_s, the relative complement of this with respect to Θ_{123}, which is $\Theta_{123} - ([\mu_s] \bigcirc \Theta_{123})$ is the set of games belonging to Θ_{123} which do not include μ_s; so Rule A gives the credal set of Luke's certainty of p as being $\Theta_{123} - ([\mu_s] \bigcirc \Theta_{123})$ which when μ_s is μ_1, is the same as $[\mu_2] \oplus [\mu_3]$. In letting Rule A regulate what are to be his fundamental credal sets in his game with Matthew, Luke will be relying on Matthew's not choosing an action inadmissible for him according to the Domination Principle, and may be said to be *having respect for Matthew's reasonableness*.

With the use of Rule A there are four fundamental credal sets

$[\mu_1] \oplus [\mu_2]$, $[\mu_1] \oplus [\mu_3]$, $[\mu_2] \oplus [\mu_3]$, $[\mu_1] \oplus [\mu_2] \oplus [\mu_3] = \Theta_{123}$, the last being given by Luke's certainty that Matthew will choose from among three possible actions. The relevant decomposition is into seven possibilities, since for every preference system V_M of Matthew's and for every game δ in which he is playing, one and only one of the following seven propositions will be true (where by an action's being admissible is meant that it is not inadmissible i.e. it is not dominated by some other action):

$g_1 = M_1$ is inadmissible and each of M_2, M_3 is admissible in δ for V_M;
g_2, g_3 similarly for M_2 and M_3 respectively;
$g_4 =$ Each of M_1, M_2 is inadmissible and M_3 is admissible in δ for V_M;
g_5, g_6 similarly for M_1, M_3 and M_2, M_3 respectively;
$g_7 =$ Each of M_1, M_2, M_3 is admissible in δ for V_M.

(It is impossible for all three M_s's to be inadmissible.)

Taking $[g_1, g_2, g_3, g_4, g_5, g_6, g_7]$ as the relevant decomposition, Rule A specifies the following seven quasifundamental credal sets, by one application in the case of g_1, g_2, g_3, by two applications in the case of g_4, g_5, g_6, by no application (as it were) in the case of g_7, Luke's certainty of which imposes by Rule A no restriction upon the complete set Θ_{123} (Numerical abbreviations are used as before for the subsets of Θ_{123}.)

$C_L(g_1), C_L(g_2), C_L(g_3) \rightarrow 236, 135, 124$
$C_L(g_4), C_L(g_5), C_L(g_6) \rightarrow 236 \cap 135 = 3, 236 \cap 124 = 2,$
$\qquad \qquad \qquad \qquad \qquad 135 \cap 124 = 1$
$C_L(g_7) \qquad \qquad \qquad \rightarrow 1234567$

If Luke takes his choice-situation as being a game with a person and takes Rule A as the unique rule regulating what are to be the family of his credal sets, this family (call it $F_A(3)$) will be the overlap closure of the conflation closure of the intersection closure of the family of the four fundamental credal sets 124, 135, 236, 1234567. $F_A(3)$ comprises 33 sets (not counting the null-set), and is exactly the same as the family $F_N(3)$ of credal sets that arise in games with nature.

However I do not think that Rule A by itself does justice to the way in which Luke will choose his action in a choice-situation with Matthew. For while it instructs him *negatively* not to allow for the possibility of

Matthew's choosing a particular action in specific circumstances, it never instructs him *positively* to allow for such a possibility. It is true that repeated applications of Rule A, as for the credal set of $C_L(g_6)$, may restrict Luke's credal set to being a one-membered set. But $[\mu_1]$ will be the credal set of $C_L(g_6)$ because it is the intersection of two credal sets each given by the rule, and not because it is directly given by the rule, as it is by the Application Postulate for games with nature. If Luke is certain of g_6, he will have to play in μ_1 not because that game is directly given by his certainty, but because if he may not allow either for the possibility of Matthew's choosing M_2 or for the possibility of Matthew's choosing M_3, all possibilities of Matthew's choosing an action other than M_1 are excluded.

So it seems reasonable to try to discover a positive rule which will supplement the negative Rule A, and the simple converse of Rule A immediately suggests itself, namely that if Luke is certain that Matthew is playing in such a game and has such a preference system that M_s is *admissible* in that game and for that preference system, then Luke must allow for the possibility of Matthew's choosing M_s, and so must himself play in a game which includes μ_s as a subgame. However the addition of its simple converse to Rule A would lead to consequences inconsistent with the postulates of the formal system in cases in which the game of Matthew's in question is a polyadic game such as $\lambda_1 \cup \lambda_2$.

For suppose that Matthew's preference system is such that there are two and only two actions of Matthew's, say M_1, M_2, that are admissible in the game in which he is playing, which is $\lambda_1 \cup \lambda_2$. Then the addition of its simple converse to Rule A requires that Luke's certainty of these propositions should reduce his credal set from the three-membered set $[\mu_1] \oplus [\mu_2] = [\mu_1, \mu_2, \mu_1 \cup \mu_2]$ prescribed by Rule A alone to the one-membered set $[\mu_1] \bigcirc [\mu_2] = [\mu_1 \cup \mu_2]$. For if Luke were to play either in μ_1 or in μ_2, he would either be not allowing for the possibility of Matthew's choosing M_2 or be not allowing for the possibility of Matthew's choosing M_1, and the converse of Rule A says that he must allow for both these possibilities. Suppose then that, in accordance with the converse of Rule A, Luke takes his credal set to be $[\mu_1 \cup \mu_2]$. If the admissibility of M_1 and of M_2 in $\lambda_1 \cup \lambda_2$ arises from the fact that Matthew's single preference structure $\mathscr{S}_1$ in L_1 prefers M_1 to M_2 whereas his single preference structure $\mathscr{S}_2$ in L_2 prefers M_2 to M_1,

Luke might well also believe that Matthew would use a stronger principle of rational choice than the Domination Principle in order to discriminate between them. To be able to use, for example, the Maximin Principle would require that Matthew's single preference structures for the separate rows should be interrelated in such a way that a simple ordering with regard to preference could be constructed for all of them together, and it would require much more than this if the stronger principle were that of maximizing expected utilities. But there is no reason why Luke should not both believe that Matthew will use the Maximin Principle and also believe sufficient additional propositions about Matthew's preference system to enable him to make the same calculations as Matthew and to arrive at the conclusion that M_1 would be rejected in favour of M_2 by the stronger principle. But M_1's being rejectable by the stronger principle is perfectly consistent with its being admissible (admissible being always used to mean admissible in accordance with the Domination Principle). So we should have the paradoxical situation that, if p_1 is the proposition that M_1 is admissible in game $\lambda_1 \cup \lambda_2$ for preference system V_M, p_2 the proposition that M_1 is rejectable by the Maximin Principle in game $\lambda_1 \cup \lambda_2$ for the suitably strengthened preference system V_M^+, p_3 the proposition that Matthew is playing in the game $\lambda_1 \cup \lambda_2$ and has V_M^+ as preference system, p_4 the proposition that Matthew will be using the Maximin Principle of rational choice to discriminate between admissible actions, Luke's credal set for $C_L(p_1 \wedge p_3)$ would be $[\mu_1 \cup \mu_2]$ but his credal set for $C_L(p_1 \wedge p_2 \wedge p_3 \wedge p_4)$ would be $[\mu_2]$. The paradox arises because, according to Postulate 6, $\Lambda^c C_L(p_1 \wedge p_2 \delta \wedge p_3 \wedge p_4) = \Lambda^c C_L(p_1 \wedge p_3) \cap \Lambda^c C_L(p_2 \wedge p_4)$, from which it immediately follows that $\Lambda^c C_L(p_1 \wedge p_2 \wedge p_3 \wedge p_4) \subseteq \Lambda^c C_L(p_1 \wedge p_3)$, whereas in our case $\Lambda^c C_L(p_1 \wedge p_2 \wedge p_3 \wedge p_4) = [\mu_2]$ is not included in $[\mu_1 \cup \mu_2] = \Lambda^c C_L(p_1 \wedge p_3)$. Since I am not prepared to abandon Postualtes 5 and 6 (for both would have to go), I require of a satisfactory rule that it should yield a set comprising both μ_1 and μ_2 in all cases in which Matthew, by using a principle of rational choice stronger than the Domination Principle, could discriminate between M_1 and M_2 when both are admissible. The simple converse of Rule A does not satisfy this requirement, and so cannot be considered as an appropriate addition to Rule A.

However, it is only in situations in which two or more actions are admissible in a polyadic game while each of them is inadmissible in at least one of the monadic games included in the polyadic game that a stronger principle of rational choice might be able to discriminate between them. If the actions are not only all admissible in $\lambda_1 \cup \lambda_2 \cup \ldots \lambda_r$ but also are all admissible in λ_1, in $\lambda_2, \ldots$ in λ_r, then they will be indifferent in each of Matthew's single preference structures $\mathscr{S}_1, \mathscr{S}_2, \ldots \mathscr{S}_r$; and it will be impossible for any principle of rational choice to discriminate between them. For it literally cannot matter in any way to Matthew which of these indifferent actions he chooses to do. So in this case a positive rule can apply, which it is convenient to state in terms of the notion of an action's being optimal.

An action will be said to be *optimal* in a game δ for preference system V_M if it is admissible in every monadic subgame of δ for V_M. It follows from the fact that, if M_s is admissible both in δ_1 and in δ_2, it is admissible in $\delta_1 \cup \delta_2$, that an action optimal in δ is optimal in every subgame of δ. In a monadic game being admissible coincides with being optimal, as it does also in all games having only one admissible action. If any action in a game is optimal, every admissible action is optimal.

The desired positive addition to Rule A can now be given by reference to an action's optimality; and Rule AB, incorporating this addition can be stated in the form that, if Luke is certain that Matthew is playing in such a game and has such a preference system that M_s is inadmissible and M_t is optimal in that game and for that preference system, then Luke *must not allow* for the possibility of Matthew's choosing M_s and *must allow* for the possibility of his choosing M_t. Luke's following Rule AB instead of Rule A will not affect Luke's credal sets when only Matthew's monadic games are concerned. But it will, for example, reduce Luke's credal set from comprising $\mu_1, \mu_2, \mu_1 \cup \mu_2$ to comprising $\mu_1 \cup \mu_2$ alone when Matthew has, and can have, no reason whatever for choosing M_1 rather than M_2 or M_2 rather than M_1. In letting Rule AB regulate what are to be his fundamental credal sets in his game with Matthew, Luke may be said to be having respect both for Matthew's reasonableness and for his freedom to make, in appropriate circumstances, a *gratuitous* choice.

This *gratuitous freedom of choice* (for I shall transpose the epithet) might equally well be called an *absolute* freedom, an *inescapable* freedom, an *unavoidable* freedom, an *ineluctible* freedom. I have (not gratuitously) chosen the adjective 'gratuitous' because an action chosen gratuitously in my sense is the interpretation I give to the famous *acute gratuit* of the existentialists. The *locus classicus* for this notion is Lafcadio's action, in André Gide's novel *Les Caves du Vatican*, in throwing his fellow-traveller, a perfect stranger to him, out of the train between Rome and Naples. In premeditating his action Lafcadio says to himself: "Un crime immotivé, quel embarras pour la police!" A gratuitous action is even more of an embarrassment to those who wish to simulate human action on a computer.

A specialized computer could be constructed, or a general-purpose digital computer programmed, to imitate Matthew's reasonableness in not choosing an inadmissible action. For this it will be necessary that there should be a built-in simulacrum of a preference system which would impose constraints upon the way in which the inputs to the computer, simulating Matthew's credal states, would produce the outputs, the computer's pseudo-actions. The computer could contain a variety of control systems at different levels, by which a possible output could be prohibited at the lowest 'Domination Principle' control, and if not so prohibited could be passed to a higher-level control e.g. a 'Maximin Principle' control. All this seems logically possible. But I cannot see how it would be logically possible either to construct a specialized computer or to programme a digital computer so that it would imitate Matthew's gratuitous freedom of choice. A computer which incorporates a randomizing device can, of course, yield outputs which individually cannot be predicted from a knowledge of its inputs and of its mode of operation. But its outputs will be probabilistically predictable from such knowledge. The basic decomposition for his basic choice-situation which Luke would have to use in playing a game with such a computer would consist of propositions each of which ascribed one of the possible chances of the output's occurring, and this would add enormously to the number of games included in his choice-situation. Luke would still, however, play his game with the computer using the Application Postulate as the rule determining his fundamental credal sets. Nor would the situation be essentially different were the computer to have incorporated in it a hierarchy of randomizing devices each randomizing the parameters used in the randomizing device next lower in the hierarchy.

For this among other reasons the ability, in appropriate circumstances, to make a choice which is gratuitous, in the specific sense which I have

BEHIND DECISION AND GAMES THEORY 45

explained, seems to me one of the essential features which distinguish a *person* from an *automaton*; and Luke will not be treating Matthew properly as a person in his game with him unless he has respect for his gratuitous freedom of choice as well as for his reasonableness. This involves his allowing Rule AB to regulate what are to be his fundamental credal sets.

If Luke, certain that Matthew is playing in game δ and has preference system V_M, is also certain that M_s is inadmissible and M_t optimal for him in δ for V_M, Rule AB prescribes that he must not allow for the possibility of Matthew's choosing M_s, i.e. the game in which Luke plays must not include μ_s as a subgame, and that he must allow for the possibility of Matthew's choosing M_t, i.e. the game in which Luke plays must include μ_t as a subgame. Applying the rule as before to the case in which $m=3$ and Luke's total game is $\sigma = \mu_1 \cup \mu_2 \cup \mu_3$, Rule AB gives the credal set of the conjunction of these certainties of Luke's as being $(\Theta_{123} - ([\mu_s] \bigcirc \Theta_{123})) \cap ([\mu_t] \bigcirc \Theta_{123})$.

The relevant decomposition for the use of Rule AB is into eleven possibilities (when the game has three columns). (This number is far less than the 56 given by combinatorial considerations because of the logical connexions between optimality and admissibility.) For every preference system V_M of Matthew's and for every game δ in which he is playing, one and only one of these eleven propositions is true:

$h_1 = M_1$ is optimal and each of M_2, M_3 is inadmissible in δ for V_M;
h_2, h_3 similarly for M_2 and M_3 respectively;
$h_4 =$ Each of M_1, M_2 is optimal and M_3 is inadmissible in δ for V_M;
h_5, h_6 similarly for M_1, M_3 and M_2, M_3 respectively;
$h_7 =$ Each of M_1, M_2, M_3 is optimal in δ for V_M;
$h_8 = M_1$ is inadmissible and each of M_2, M_3 is admissible but not optimal in δ for V_M;
h_9, h_{10} similarly for M_2 and M_3 respectively;
$h_{11} =$ Each of M_1, M_2, M_3 is admissible but not optimal in δ for V_M;

None of the four propositions h_8, h_9, h_{10}, h_{11} can be true if δ is a monadic game. Taking $[h_1, h_2, \ldots h_{11}]$ as the relevant decomposition, the eleven quasifundamental sets are specified as follows, Rule AB being used to give the credal sets in the cases of h_1 to h_7, Rule A alone being used in the cases h_8, h_9, h_{10} when the B-part of Rule AB does not apply, neither rule being used in the case of h_{11}, where neither rule applies and when no restriction is imposed upon the complete set Θ_{123}. (Numerical abbreviations are used as before.)

$C_L(h_1), C_L(h_2), C_L(h_3) \rightarrow 1457 \cap 135 \cap 124 = 1,$
$\qquad 2467 \cap 236 \cap 124 = 2, 3567 \cap 236 \cap 135 = 3$
$C_L(h_4), C_L(h_5), C_L(h_6) \rightarrow 1457 \cap 2467 \cap 124 = 4,$
$\qquad 1457 \cap 3567 \cap 135 = 5, 2467 \cap 3567 \cap 236 = 6$
$C_L(h_7) \qquad\qquad\qquad \rightarrow 1457 \cap 2467 \cap 3567 = 7$
$C_L(h_8), C_L(h_9), C_L(h_{10}) \rightarrow 236, 135, 124$
$C_L(h_{11}) \qquad\qquad\qquad \rightarrow 1234567$

If Luke takes his choice-situation as being a game with a person and accepts Rule AB as the appropriate rule for such a game, the family of his fundamental credal sets will be [1457, 2467, 3567, 124, 135, 236, 1234567] and the family of his credal sets (call this family $F_{AB}(3)$) will be the overlap closure of the conflation closure of the intersection closure of this family of seven fundamental credal sets. $F_{AB}(3)$ comprises 60 sets, (not counting the null-set) including the 33 sets of $F_A(3) = F_N(3)$, and (as can easily be proved) is the same as the family $P(3)$, of the non-empty perfect subsets of Θ_{123}. (Unlike $F_N(m) = F_A(m)$, I have discovered no interesting lattice-theoretic characterization of $P(m) = F_{AB}(m)$.)

Before commenting on this surprising result, I will complete the record by saying what will happen if Luke takes his choice-situation as being a game with a person and accepts the positive part B of Rule AB without accepting the negative Rule A. If Luke respects Matthew's gratuitous freedom of choice but has no respect for his reasonableness, the quasi-fundamental credal sets will be [1457, 2467, 3567, 47, 57, 67, 7, 1234567]; and the family $F_B(3)$ constructed out of them by the overlap closure of the conflation closure of an intersection closure will comprise 18 sets, and will be D_7, the free distributive lattice with 1457, 2467, 3567 as generators. For every m, $F_B(m) \subseteq F_A(m) \subseteq F_{AB}(m)$; and when $m \geqslant 3$ these inclusions are proper inclusions, i.e. $F_B(m) \subset F_A(m) \subset F_{AB}(m)$.

The numbers of sets in the various families are given in Table I for $m = 1, 2, 3, 4$. $D^*(m)$ denotes the free distributive lattice with the m generators $[\mu_s] \bigcirc \Theta_{12...m}$, $s = 1, 2,...m$. $F_\Theta(m)$ denotes the family of all the non-empty subsets of $\Theta_{12...m}$: the number of these is $2^{2^{m-1}} - 1$. Complicated calculations are required to obtain the larger numbers in the first and third columns: the numbers in the first column were first calculated by Dedekind in 1897.

TABLE I
Number of sets in the families

	$F_B(m) = D^*(m)$	$F_N(m) = F_A(m)$	$F_{AB}(m) = P(m)$	$F_\Theta(m)$
$m = 1$	1	1	1	1
2	4	6	6	7
3	18	33	60	127
4	166	266	2,479	32,767

Table I shows that, with increasing m, the numbers in the third column separate very rapidly from those in the second. When $m = 3$, there are fewer sets of F_{AB} which do not belong to F_N than of those which do so belong; when $m = 4$, there are more than eight times as many. The table also shows why my argument required that we should consider games with at least three columns, since $F_N(2) = F_A(2) = F_{AB}(2)$.

I must now comment upon the surprising result of my formal theory, that the family of credal sets that can appear in a game of Luke's with Matthew, if Luke respects both Matthew's reasonableness and his gratuitous freedom of choice, is larger than the family that can appear in a game of Luke's with nature. Indeed it is as large as a family of credal sets (of games with the same number of columns) can possibly be, since the family contains all non-empty perfect sets and every credal set is perfect.

The 27 sets which belong to $F_{AB}(3)$ but which do not belong to $F_N(3)$ all arise because of the four quasifundamental credal sets 4, 5, 6, 7 which are not fundamental credal sets in a game with nature. These four sets all appear as credal sets of certainties that two actions (in the case of $4 = [\mu_1 \cup \mu_2]$, 5, 6) or all three actions (in the case of $7 = [\mu_1 \cup \mu_2 \cup \mu_3]$) are optimal in the game δ in which Matthew is playing and for his preference system V_M. But two actions, e.g. M_1, M_2, can only both be optimal in δ if in Matthew's single preference structures in every row of δ, M_1 is neither preferred to M_2, nor M_2 to M_1, so that he has a gratuitous freedom of choice between them. If Matthew's preference system is such that none of his single preference structures $\mathscr{S}_1, \mathscr{S}_2, \ldots \mathscr{S}_1$

contain any indifferences (i.e. absences of preference one way or the other), he will have no occasion to make a gratuitous choice; and Luke, if he is certain that Matthew's preference system fulfils this condition, will never use the quasifundamental credal sets 4, 5, 6, 7 yielded by certainties of h_4, h_5, h_6, h_7 respectively in constructing his credal sets, the family of which will then be the same as $F_A(3) = F_N(3)$. Thus the 27 extra sets of $F_{AB}(3)$ arise as a consequence of Luke's respecting Matthew's gratuitous freedom of choice in circumstances in which he lacks the certainty that Matthew's preference system, by containing no indifferences, will never give him an opportunity to exercise this freedom.

A simple example of a game of Luke's with Matthew which is yet complex enough to illustrate all the problems is the game in which Matthew has to choose between three possible actions (columns M_1, M_2, M_3) and Luke to choose between two (rows L_1, L_2), and when Luke is certain that Matthew's single preference structures $\mathscr{S}_1, \mathscr{S}_2$ in L_1, L_2 respectively (the relevant parts V_M of his preference system) are as follows:

$\mathscr{S}_1$. (In L_1). M_2 is preferred both to M_1 and to M_3, and M_3 is preferred to M_1;

$\mathscr{S}_2$. (In L_2). Both M_1 and M_3 are preferred to M_2, but neither M_1 nor M_3 is preferred to the other.

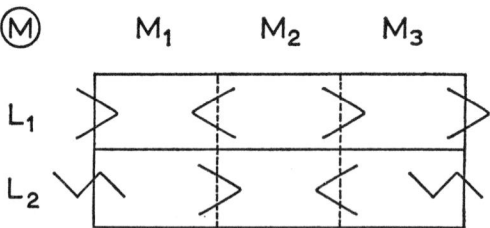

If Matthew has V_M as his preference system, then in his game λ_1, M_2 is optimal and M_1, M_3 are inadmissible, i.e. h_2; in his game λ_2, M_1, M_3 are optimal and M_2 is inadmissible, i.e. h_5; in his game $\lambda_1 \cup \lambda_2, M_1$ is inadmissible (since it is dominated by M_3) and M_2, M_3 are admissible but not optimal i.e. h_8. So, since Luke is taking for granted that Matthew must play in one and in only one of these three games, the relevant decomposition reduces to $[h_2, h_5, h_8]$.

For $\Lambda^c C_L(h_2)$, and thus also for $\Lambda^c C_L$ (Matthew has preference system V_M and is playing in λ_1) – abbreviated into $\Lambda^c C_L\{V_M; M \text{ in } \lambda_1\}$ – both Rule A and Rule AB produce $[\mu_2] = 2$, Rule A because $236 \cap 124 = 2$, Rule AB because $2467 \cap 236 \cap 124 = 2$.

For $\Lambda^c C_L(h_5)$, and thus also for $\Lambda^c C_L\{V_M; M \text{ in } \lambda_2\}$, Rule A produces $[\mu_1] \oplus [\mu_3] = 135$, but Rule AB produces $1457 \cap 3567 \cap 135 = 5 = [\mu_1 \cup \mu_2]$.

For $\Lambda^c C_L(h_8)$, and thus also for $\Lambda^c C_L\{V_M; M \text{ in } \lambda_1 \cup \lambda_2\}$, both Rule A and Rule AB produce $[\mu_2] \oplus [\mu_3] = 236$.

If Luke goes by Rule A, the quasifundamental credal sets are 2, 135, 236; and the seven additional credal sets which can be obtained from them by conflation closure followed by overlap closure are 26, 467, 2467, 34567, 134567, 234567, 1234567. All these belong to $F_N(3)$.

If Luke goes by Rule AB, the quasifundamental credal sets are 2, 5, 236: these all belong to $F_N(3)$. But of the eight additional credal sets that can be obtained from the family [2, 5, 236] by conflation closure followed by overlap closure, namely 7, 26, 57, 27, 257, 267, 2567, 23567, only the first three belong to $F_N(3)$, the remaining five being sets belonging to $P(3)$ which do not belong to $F_N(3)$.

The largest of the five new sets, 23567, is the credal set of several of Luke's creeds of which the simplest is

$$C_L\{V_M; M \text{ in } \lambda_1 \vee M \text{ in } \lambda_2 \vee M \text{ in } \lambda_1 \cup \lambda_2\}.$$

Since the alternation is a logically necessary proposition, this reduces to $C_L\{V_M\}$ – to use an obvious modification of my abbreviation. So the other creeds whose credal set is 23567 impose no limitation upon Θ_{123} beyond that already imposed by Luke's certainty that Matthew has preference system V_M.

The next largest, 2567, I mention because I shall take it as the example of Section III. It is the credal set of several creeds of which the simplest is

$$C_L\{V_M\} \wedge U_L\{M \text{ in } \lambda_1 \cup \lambda_2, M \text{ in } \lambda_1 \vee M \text{ in } \lambda_2\}.$$

The uncertainty of Luke's in this conjunction of creeds might derive from his uncertainty as to whether or not Matthew believed that Luke's preference system was such as to leave Luke, of whom Matthew would be certain that he would use or invent some principle of rational choice to prescribe his choice between L_1 and L_2 if he could possibly do so, no option but to make a gratuitous choice between L_1 and L_2.

To conclude this part. It will have been noticed that, if Luke is certain that Matthew's preference system is V_M and that he is playing in the game λ_2 in which M_1, M_3 are optimal and M_2 inadmissible (the case of h_5), Luke's credal set will be $[\mu_1 \cup \mu_3]$, which is exactly the same as it would be if he were playing a game with nature and were uncertain between two propositions m_1, m_3. So Luke's certainty about Matthew's preference system etc. will guide his choice of action in exactly the same way *as if* he were uncertain whether Matthew would choose M_1 or would choose M_3. But there is no objection to Luke's being uncertain about this and still be playing a genuine game with Matthew, provided that the explanation he would give of $[\mu_1 \cup \mu_3]$ being the credal set that guided his choice of action was not only that he was uncertain as to which action Matthew would choose, but also that his uncertainty was based upon a certainty that Matthew was in a situation where he would have to make a gratuitous choice between the two actions. In a game with a randomizing automaton an uncertainty can be resolved by guessing the method of randomization and putting the guessed chances into a new basic decomposition describing the choice-situation: in a game with Matthew there is no way of resolving an uncertainty arising from Matthew's gratuitous freedom of choice except by giving up the assumption that Matthew's preference system is what Luke has been taking it to be, one that necessitates Matthew's making a gratuitous choice.

III

I shall explain how an observer, Karl, of Luke's behaviour can refute the hypothesis that Luke's creed is such that, in a specific choice-situation, its credal set is a specific set Γ by taking one example only – that in which $\Gamma = [\mu_2, \mu_1 \cup \mu_3, \mu_2 \cup \mu_3, \mu_1 \cup \mu_2 \cup \mu_3] = 2567$ in a choice-situation $\sigma = \mu_1 \cup \mu_2 \cup \mu_3$ in which there are three columns. The method I shall give can be applied to any Γ and to choice-situations with any number of columns, but in order that the refutation should be discriminative (i.e. that the refutation of distinct Γ's should be different) it is necessary, in most of the cases in which Γ is a credal set which cannot occur in games with nature, for the choice-situation of Luke's concerned in the refutation to have a number of rows greater than the number of columns.

The requirement raises the difficulty that, in a game with a person, the number l of rows from which Luke has to choose must be equal to the number of λ's in Matthew's choice-situation $T = \lambda_1 \cup \lambda_2 \cup \ldots \lambda_l$, so that, in our example of Luke's 2×3 game with Matthew, Luke's choice-situation cannot have the four rows required for the discriminative refutation of Luke's credal set being 2567 without Matthew's choice-situation being different from the $T = \lambda_1 \cup \lambda_2$ it was taken to be. However this difficulty can be overcome by regarding $T = \lambda_1 \cup \lambda_2$ as $T^* = \lambda_{11} \cup \lambda_{12} \cup \lambda_{21} \cup \lambda_{22}$ with Matthew's preference system V_M^+ for T^* having single preference structures $\mathscr{S}_{11}$, $\mathscr{S}_{12}$ for λ_{11}, λ_{12} respectively which are the same as the single preference structure $\mathscr{S}_1$ for λ_1 and similarly single preference structures $\mathscr{S}_{21}$, $\mathscr{S}_{22}$ which are the same as $\mathscr{S}_2$. Then if, as in our example, M_2 is optimal and M_1, M_3 inadmissible in λ_1 for V_M, M_2 will be optimal and M_1, M_3 inadmissible in λ_{11}, in λ_{12}, in $\lambda_{11} \cup \lambda_{12}$ for V_M, and similarly with the other cases, so that the relevant decomposition will still be $[h_2, h_5, h_8]$ and the quasifundamental credal sets will still be 2, 5, 236. The situation may be thought of as one in which two of Luke's actions L_{11}, L_{12} may be regarded as two alternative specific ways of doing the same generic action L_1, and L_{21}, L_{22} regarded as two alternative specific ways of doing the same generic action L_2, when it makes no difference to Matthew which of L_{11}, L_{12} or which of L_{21}, L_{22} Luke performs.

So for the testing of $\Gamma = 2567$ Luke's game with Matthew will be taken as being a 4×3 game, with Luke's four possible actions renumbered (in any order) as L_1, L_2, L_3, L_4.

Karl will be assumed to be certain that the choice-situation is a 4×3 one, that Luke is reasonable enough never to choose a row which is inadmissible (i.e. dominated by another row) and that Luke has a specified and different preference system on two occasions of his choosing in the same choice-situation, his credal set being the same on each occasion. These are big assumptions to make, but nothing can be done without them.

Consider two of Luke's possible preference systems which will be called $V_L(24567)$, $V_L(123567)$. They can be conveniently represented by arrays in which Luke's single preference structure in each column is represented by the greater or less or equal relationships between the numbers in the cells of the column.

Ⓛ	M_1	M_2	M_3
V (24567) is L_1	1	1	1
L_2	1	1	1
L_3	0	0	1
L_4	1	0	0

i.e. in $\mathscr{R}_1$ none of L_1, L_2, L_4 is preferred to any other, and each of L_1, L_2, L_4 is preferred to L_3; and similarly for the other columns.

Let L_r is inadmissible *throughout* Ξ be an abbreviation for L_r is inadmissible in every game belonging to Ξ and is admissible in every game belonging to Θ_{123} but not belonging to Ξ, i.e. in every game belonging to $\Theta_{123} - \Xi$. For $V_L(24567)$ L_3 is inadmissible throughout 124567 and L_4 is inadmissible throughout 234567.

Ⓛ	M_1	M_2	M_3
V_L (123567) is L_1	1	1	1
L_2	1	1	2
L_3	2	0	0
L_4	0	2	0

For V_L (123567) L_1 is inadmissible throughout 123567.

If Luke chooses L_3 while having as preference system $V_L(24567)$, his credal set Λ cannot be included in 124567; for, if it were, the game in which he played would belong to 124567, in every game of which L_3 is inadmissible.

Similarly if Luke chooses L_4 while having his preference system $V_L(24567)$, Λ cannot be included in 234567; and if Luke chooses L_1 while having as preference system $V_L(123567)$, Λ cannot be included in 123567.

Let p be the proposition that on one occasion of the choice-situation when Luke's credal set was Λ and his preference system was $V_L(24567)$ Luke performed action L_3, q be the proposition that on the same occasion of the choice-situation Luke performed action L_4, r be the proposition that on another occasion of the choice-situation when Luke's credal set was also Λ but his preference system was $V_L(123567)$ Luke performed action L_1.

Then it logically follows from $p \vee q \vee r$ that

$$\text{Not } (\Lambda \subseteq 124567) \vee \text{Not } (\Lambda \subseteq 234567) \vee \text{Not } (\Lambda \subseteq 123567),$$

which is equivalent to

$$\text{Not } ((\Lambda \subseteq 124567) \wedge (\Lambda \subseteq 234567) \wedge (\Lambda \subseteq 123567)),$$

i.e. to

$$\text{Not } (\Lambda \subseteq 124567 \cap 234567 \cap 123567),$$

i.e. to

$$\text{Not } (\Lambda \subseteq 2567).$$

If Karl believes the alternation of propositions which is the protasis of this implication, he can deduce that Luke's credal set was not included in 2567. But Karl's belief in the protasis will not enable him to deduce, for any set Ξ not included in 2567, that Λ was not included in Ξ. Acceptance of the protasis will therefore pick out 2567 as the union of all those sets which are proved by the alternation of three facts about Luke's behaviour not to include Λ; and belief in the protasis will thus refute the hypothesis that Λ is included in 2567 without refuting any hypothesis of which this hypothesis is not a logical consequence. The hypothesis $\Lambda \subseteq 2567$ will be *the weakest* hypothesis refutable in this way.

The two preference systems called $V_L(24567)$, $V_L(123567)$ used in this testing procedure were chosen so as to indicate how the procedure can be applied to every set Γ included in $\Theta_{12...m}$, for any m. Each of the sets belonging to the family $F_N(m)$ can be tested by using one preference system constructed in each case according to a general rule; and each of the sets $\Theta_{12...m} - [\gamma]$, where γ is a game belonging to $\Theta_{12...m}$ can also be tested by using one preference system constructed in accordance with a general rule. So if Γ_1 is the smallest set belonging to $F_N(m)$ which includes Γ, then either $\Gamma = \Gamma_1$, in which case only one preference system is required for the test, or $\Gamma_1 - \Gamma = [\gamma_1] \cup [\gamma_2] \cup ... [\gamma_s]$, where the γ's are distinct games belonging to $\Theta_{12...m}$. In this latter case Not $(\Lambda \subseteq \Gamma)$ is equivalent to

Not $(\Lambda \subseteq \Gamma_1 \cap (\Theta_{12...m} - [\gamma_1]) \cap (\Theta_{12...m} - [\gamma_2]) \cap ... (\Theta_{12...m} - [\gamma_s]))$,

i.e. to

Not $(\Lambda \subseteq \Gamma_1) \vee$ Not $(\Lambda \subseteq \Theta_{12...m} - [\gamma_1]) \vee ...$ Not $(\Lambda \subseteq \Theta_{12...m} - [\gamma_s])$.

So $1 + s$ preference systems constructed in accordance with general rules are required for the testing procedure. In our case $2567 = 24567 - 4 = 24567 \cap (\Theta_{123} - 4) = 24567 \cap 123567$, and two preference systems are used.

For some subsets of $\Theta_{12...m}$ which do not belong to $F_N(m)$, and indeed for all subsets of Θ_{123}, a single preference system can be found to provide the test. For our case of $\Gamma = 2567$ this system is

Ⓛ	M_1	M_2	M_3
L_1	1	0	1
L_2	0	1	1
L_3	1	0	0
L_4	0	0	1

For this preference system L_3 is inadmissible throughout 23567 and L_4 is inadmissible throughout 124567; so Luke's either choosing L_3 or choosing L_4 while having this preference system will refute the hypothesis $\Lambda \subseteq 23567 \cap 124567 = 2567$.

A creed whose credal set is 2567 will be weaker for Luke than a creed whose credal set, e.g. 257, is properly included in 2567 in the sense that, by imposing less of a restriction upon the set of games he may play in, it will provide Luke with less guidance in his choice of action. So, whereas a refutation of $\Lambda \subseteq 2567$ will refute a proposition that Luke held a creed q_1 which is such that $\Lambda^c q_1 \subset 2567$ as well as a proposition that Luke held a creed q_2 which is such that $\Lambda^c q_2 = 2567$, the latter creed will be weaker than the former. So the hypothesis that Luke held a creed whose credal set is 2567 can be picked out as *the weakest* hypothesis about Luke's creed that is refutable by Karl's knowledge of three facts about Luke's behaviour.

It will be remembered that 2567 is one of the credal sets that can only occur in a game with a person. The hypothesis that Luke was playing in such a game (and with a creed whose credal set is 2567) can be refuted by knowledge of his preference systems and behaviour on two occasions of his playing the game. This method of discriminative falsification is my contribution towards furthering the pragmatists' and verificationalists' programme of taking how a man acts as the criterion for what he believes.

ADDED IN PROOF MARCH 1972

An expanded and improved version of the argument of this paper will appear in my book *Pure Theory of Applied Belief* to be published by the Cambridge University Press. In this book the single preference structures will be explicated in terms of certainties of conditional propositions; and a simple modal postulate system, interpreted as a doxastic logic, will take the place of the postulates about credal sets of games.

ISAAC LEVI

COMMENTS

Professor Braithwaite's aim is to examine in a systematic way inferences which an observer, Karl, can make concerning Luke's beliefs from information about Luke's choices from among alternative courses of actions.

Braithwaite explicitly assumes that in addition to knowing Luke's choice, Karl has other information about Luke at his disposal. In particular, Karl knows that Luke considers himself as having a choice among the options $l_1, l_2, ..., l_m$, that Luke is certain that at least and at most one of the propositions $p_1, p_2, ..., p_n$ belonging to the set U is true, and that, given any p_i in U, Luke has a preference ranking among the l_j's on the assumption that p_i is true. That ranking is representable by the 'column' M_i.

To reach some conclusion about Luke's beliefs from Luke's choice among the l_j's, Karl must know how Luke's beliefs contribute to determining his choice. That is to say, Karl must have a theory about Luke's way of determining choices from preferences and beliefs. Professor Braithwaite introduces such a theory – one which he would consider as plausibly applicable to Luke when Luke is rational. Part of that theory conforms fairly well to conventional wisdom in decision theory. Part of the theory does not. I claim that the part of the theory that deviates from orthodox approaches has nothing to recommend it either as decision theory or as a device for enabling Karl to reach conclusions about Luke's beliefs. Yet, I think that Professor Braithwaite has contributed something of interest to an account of how beliefs can be inferred from choices.

In explaining my meaning, I shall restrict my discussion to cases where Luke is playing a 'game against nature'. The points I wish to make carry over mutatis mutandis to the two person game quite nicely.

Professor Braithwaite's theory of how beliefs determine choices regards decision making as a three stage process.

At stage 1, the agent, Luke, fixes his beliefs and thereby determines a 'credal set' whose members are games from which Luke must choose one to play.

At stage 2, Luke chooses a game from his credal set in a manner which Braithwaite leaves unexplained. We shall return to this point subsequently.

At stage 3, Luke chooses an option l_i from his set of options such that given the game which Luke is playing and the set of options, l_i is admissable – i.e., is not dominated by any other option in that game.

To fix ideas, suppose that Luke has two options l_1 and l_2. U consists of three propositions p, q and r. Let M_1, M_2 and M_3 be associated with p, q, and r respectively. Finally, let l_1 and l_2 be ranked equally in M_1 and M_2 and l_1 be preferred to l_2 in M_3.

There are many cases which one might consider. I will consider only three; but they should illustrate the general points.

Case 1: Luke is certain of p. According to Braithwaite, stage 1 and stage 2 collapse here; for Luke's creed determines a credal set whose only member is the game (M_1). Hence, Luke's creed determines a unique game for him to play. At stage 3, Luke may choose either l_1 or l_2 since either option is admissible in (M_1).

Case 2: Luke is uncertain between p and q. According to Braithwaite, Luke's creed determines a single membered credal set here as well. The game is (M_1, M_2). Once more, stages 1 and 2 collapse. At stage 3, Luke may choose either l_1 or l_2 since both options are admissible in (M_1, M_2).

Case 3: Luke is certain of $p \vee q$ but is not certain of p and is not certain of q. We may describe the situation as follows: If W is the set of propositions which are disjunctions of elements of U, $p \vee q$ is a strongest proposition in W of which Luke is certain. According to Braithwaite, Luke's creed determines a three membered credal set consisting of the games (M_1), (M_2), (M_1, M_2). In each of these three games both options are admissable.

Before turning to what Karl can learn from Luke's choice about Luke's creed, I wish to consider what I believe to be a defect in the theory – namely concerning Luke's choice of a game to play in situations such as case 3.

If case 3 were the only case to consider the matter would be of small importance; for both options are admissable in all games in the credal set.

Consider, however, case 4 where the strongest proposition in W of which Luke is certain is $p \vee r$. On Braithwaite's view, the credal set consists of (M_1), (M_3) and (M_1, M_3). l_1 is admissible in all three games. l_2 is admissible only in (M_1). Thus, Luke may choose l_2 only if he selects (M_1) at stage 2 as the game to play at stage 3.[1]

Although Braithwaite introduces no procedure for determining choice among games at stage 2, there is an obvious and quite plausible way of doing so. If Luke is a rational man, he should take into consideration all possibilities which he has not definitely ruled out – i.e., he should include in the game he plays a column for each proposition in U which he is not convinced is false. Luke would be foolish to ignore any possibility unless he were certain it would not obtain.

If one endorses this principle, in case 3, Luke is obliged to choose the game (M_1, M_2). For all practical purposes case 3 collapses into case 2. Braithwaite distinguishes between the two cases on the grounds that he discerns two kinds of creeds which determine different credal sets. However, once one introduces the principle I have just suggested, both kinds of creeds, in effect, determine the same game for Luke to play.

This example illustrates a peculiarity of Braithwaite's theory. Braithwaite needs a principle for choosing among games at stage 2. Once a plausible principle is specified, however, there is no longer any need for identifying a special intermediate stage between fixing beliefs and choosing policies. Once beliefs are fixed, games are uniquely determined. All that is left is to choose among the options.

This seems to constitute a strong case for dropping the complicated proposal introduced by Braithwaite which includes mystifying distinctions between uncertainty between p and q and absence of certainty of p and of q. It is enough to characterize Luke's creed in terms of the strongest proposition in W of which Luke is certain. That determines a single game for Luke to play. Case 2 and case 3 collapse into one.

Thus far, nothing has been said about the chief problem considered by Braithwaite – to wit, how to make inferences about Luke's beliefs from his choices. It is conceivable that the hithertofore invisible virtues of Braithwaite's three stage theory of belief and action might shine forth in this context. Such is not the case.

Suppose that Karl learns that Luke chooses l_1. What conclusions may Karl draw about Luke's beliefs?

The following is a restatement of Braithwaite's analysis of the situation:
(1) l_1 is admissable in (M_1), (M_2), (M_3), (M_1, M_2), (M_1, M_3), (M_2, M_3), (M_1, M_2, M_3).
(2) Hence, as far as Karl knows, the game which Luke chose to play at stage 2 could have been any one of these seven games.
(3) The 'credal set' of games from which Luke chose at stage 2 is the set consisting of all seven games.
(4) That credal set is the one determined by the creed of being uncertain between p, q and r.
(5) Hence, Karl should conclude that Luke is uncertain between p, q and r.

This series of statements captures, insofar as I can make out, the sort of reasoning that Karl would be obliged to follow to reach the conclusion recommended by Braithwaite when Karl knows that Luke chooses l_1. The argument, however, is fallacious. Before showing that it is fallacious, we shall consider the analysis of how Karl should reason if one follows the orthodox two stage view of decision and regard Luke's game as uniquely determined by the strongest proposition in W of which he is certain.
(1) l_1 is admissible in (M_1), (M_2), (M_3), (M_1, M_2), (M_1, M_3), (M_2, M_3), (M_1, M_2, M_3).
(2) Hence, the game which Luke is playing is one of these seven games.
(3) Hence, the strongest proposition of which Luke is certain is either $p, q, r, p \vee q, p \vee r, q \vee r$, or $p \vee q \vee r$.

Notice that the orthodox two stage theory proceeds along lines analogous to Braithwaite's three stage theory for two steps. By using the assumption that Luke chooses only admissible options, both theories allow substantially the same inference concerning the game which Luke is playing. The difference between the two theories arises because of differences in the way in which the two theories regard that game as being determined by Luke's creed.

The two stage theory precludes inferring that Luke is in a definite credal state save in situations where Luke's choice rules out all but one game as the game played by Luke. Braithwaite's theory seems (but only seems) to have an advantage here. Karl, it would appear, is in a position to conclude something definite about Luke's credal stage. Appearances, however, are deceiving.

Consider, once more Braithwaite's analysis. Step 3 of the argument contends that the credal set – i.e., the set of games determined by Luke's creed at stage 1, is the seven membered set. This assumption entitles one to conclude via step 4 that Luke is uncertain between p, q and r.

Step 3 is fallaciously obtained. It is true that as far as Karl knows Luke based his choice at stage 3 on any one of the seven games listed. That is all Karl knows. From this nothing follows concerning the nature of the credal set from which Luke chose at stage 2 the game he played at stage 3. Hence, nothing follows concerning Luke's creed which determined at stage 1 Luke's credal set.

Thus, in spite of appearances to the contrary, adoption of Braithwaite's theory of the relation between belief and choice does not permit Karl to make inferences from Luke's choices to Luke's beliefs which are any more definite than those allowed by the orthodox theory. Braithwaite's three stage theory has little to recommend it. It is unnecessarily complex both for purposes of explanation and purposes of reaching conclusions about beliefs from information about choice. I think it should be abandoned.

Rejection of Braithwaite's three stage theory of the relation between belief and choice ought not, however, lead to a dismissal of his theory of credal sets. The theory is of considerable interest provided that it is properly interpreted in terms of an appropriate theory of belief and action. The point to notice is that when the orthodox two stage theory is adopted, Braithwaite's credal sets are no longer credal in Braithwaite's sense – i.e., Luke's credal set is no longer a set of games determined by Luke's creed from which Luke is free to choose a game to play at stage 3. In this sense of 'credal', all credal sets are one-membered.

What Braithwaite regards as a credal set is not determined by Luke's beliefs but by Karl's beliefs concerning the game which Luke plays. In effect, it represents Karl's creed about Luke's creed. Thus, when Karl knows that Luke chose l_1, he knows that Luke was playing one of seven games and, hence, that one of seven propositions is the strongest proposition in W of which Luke is certain. If Karl knows that Luke chose l_2, he knows that Luke is playing one of three games (M_1), (M_2) or (M_1, M_2) and, hence, Karl knows that the strongest proposition in W of which Luke is certain is either p, q or $p \vee q$ without knowing which of these three propositions it is.

The interesting fact is that Braithwaite's theory of credal sets may be interpreted in this way. The whole mathematical characterization of credal sets holds under the new interpretation. Since the theory so interpreted does contribute to an understanding of the range of inferences concerning Luke's beliefs which may be drawn by Karl from knowledge of Luke's choices, it is relevant to Braithwaite's problem.

To sum up, Braithwaite's three stage theory of the relation between belief and choice has no advantages and some liabilities as compared to the orthodox two stage view. Yet, his theory of credal sets when appropriately reinterpreted in terms of the two stage view does contribute to our understanding of how knowledge of belief can be obtained from knowledge of choice.

Columbia University

NOTE

[1] The matter would have been still more vital, had l_2 been preferred to l_1 in M_1. In that event, both options are admissable in (M_1, M_3), l_1 is admissable in (M_3) and l_2 in (M_1).

RONALD GIERE

COMMENTS

Braithwaite's paper, which was delivered with great charm and vigor at the conference, has two notable objectives. The primary objective, in the words of the final sentence, is to contribute "toward furthering the pragmatists' and verificationists' program of taking how a man acts as the criterion for what he believes." (p. 55) The second objective, which emerges almost as a by-product of the analysis, is to point out "one of the essential features which distinguish a *person* from an *automaton*." (p. 45) I will comment briefly on both objectives.

Philosophers interested in relations between belief and action have not found game theory as helpful as was once hoped because game theory, as such, provides no means for separating the roles of beliefs and values in the choice of actions. Braithwaite's goal is to provide an independent role for beliefs by treating an ordinary choice situation as but a segment of a larger *Ur*-choice situation. Thus the game treated by decision theorists is thought of as a subgame of a larger *Ur*-game. Braithwaite hopes to show that the selection of subgames from the *Ur*-game is a function of the agents beliefs alone. Values would then come in only at the second stage when a single action is chosen.

An immediate problem with this general program is that the elements of the *Ur*-choice situation, like those of any choice situation, can hardly be supposed to be determined solely by pure logic or even by physical laws. Beliefs and values are already involved. The choice of the set of alternative actions taken to be open to the agent, for example, must already reflect both his beliefs and values. Thus the most Braithwaite could expect from his theory is to be able to isolate *some* beliefs *relative* to a background of undifferentiated beliefs and values.

In Section III of the paper, Braithwaite shows how an observer, Karl, could go about refuting the hypothesis that an agent, Luke, is restricting himself to certain subgames, and thus refute some hypotheses as to Luke's beliefs since Luke's 'credal set' of subgames is determined by his beliefs ('creeds'). But as Braithwaite acknowledges, in order to falsify a hypothesis about Luke's credal set, Karl must already have knowledge of Luke's credal set and value system independently on one or more occasions when Luke actually chooses a particular action. It is questionable whether the necessity for such assumptions is compatible with the pragmatist-verificationist program of using action as a 'criterion' of belief. If determining even the negation of a hypothesis about Luke's belief state independently of his values requires independent assumptions concerning others of Luke's beliefs and values taken separately, it is only in a very weak sense that action is serving as a 'criterion' of belief. Perhaps Braithwaite would wish to argue independently for the appropriateness of this weakened connection. Perhaps he would argue that to attempt even only falsification of belief hypotheses without assuming some independently ascertained beliefs and values would be to attempt an undesirable 'reduction' of belief to action. At any rate, more could be said concerning these broader aspects of the whole program.

Examining Braithwaite's attempt to distinguish games with a person from games with nature or with an automaton requires further detailed knowledge of his theory. Considering games with nature first, let the *Ur*-game consist of l rows, Luke's possible actions, and m columns, the possible states of nature. Taking p and q to represent conjunctions or disjunctions of propositions referring to the columns, $M_1, M_2, ..., M_m$, Luke's relevant beliefs, or creeds, are restricted to two basic forms: Luke is certain of p, $C_L(p)$; and, Luke is uncertain between p and q, $U_L(p, q)$. Conjunctions of creeds are also creeds. Now Braithwaite presents a set of postulates according to which any creed determines a set of subgames of the *Ur*-game, the credal set. He does this, in effect, by reducing any creed to conjunctions of the form $C_L(m_i)$ and employing the *application* postulate that this creed selects the single subgame, μ_i, consisting of the l rows and the single column M_i. For the purposes of this discussion, only the fifth postulate requires further mention. Suppose the total game consists of two rows and three columns, M_1, M_2, M_3. If Luke's creed is $C_L(m_1 \vee m_2)$, then according to Postulate 5, Luke's credal set is $[\mu_1, \mu_2, \mu_1 \cup \mu_2]$, i.e., a set consisting of three of the seven basic subgames of the *Ur*-game. In Braithwaite's scheme, this set is the 'conflation' of the sets corresponding to the two games with single columns M_1 and M_2 respectively. Braithwaite calculates that there are 33 different sets of games that could belong to some credal set and shows how to attain this family by applying the operations of conflation and overlap to the set consisting of the 3 monadic games alone.

Imagine next that the states of nature are in fact the result of possible actions of a second player, Matthew. If Luke is to take account of Matthew, the scope of his creeds must be expanded, and Braithwaite expands it to include Luke's being certain of Matthew's value system, V_m. Braithwaite then considers the application rule, Rule A, which simply instructs Luke *not* to include in his credal set any game including a column representing an action which for Matthew is inadmissible, i.e., dominated by some other action. Braithwaite calls this having respect for Matthew's reasonableness. However, taking the 2×3 game again, we see that Luke's credal set is the same whether he is playing a game with Matthew in which $C_L(V_M)$ and M_3 is dominated or he is playing with nature and $C_L(m_1 \vee m_2)$. This result holds in general. The family of possible credal sets in these cases contains the same 33 members as that Braithwaite calculated for games with nature.

At this point Braithwaite complains that Rule A only instructs Luke *negatively* to ignore certain of Matthew's possible actions – it never instructs him *positively* to allow for certain actions. This complaint is misleading in that Rule A does tacitly instruct one positively to keep all members of the total credal set not eliminated by the dominance principle. Braithwaite nevertheless proposes an addition to Rule A which requires that all subgames in Luke's credal set *include* any optimal action of Matthew's. An optimal action is one admissible in every subgame, and two or more actions can be optimal if they have the same values throughout. Consider now the case in which of the three initial actions, M_1, M_2, M_3, is dominated while M_1 and M_2 are optimal. Rule A says that the credal set is $[\mu_1, \mu_2, \mu_1 \cup \mu_2]$ while Rule AB dictates $[\mu_1 \cup \mu_2]$. So Rule A and Rule AB yield different credal sets. Indeed, Braithwaite goes on to show that there are 60 members of the family of possible credal sets under Rule AB including the 33 found in games with nature.

According to Braithwaite, Rule AB allows Luke to respect both Matthew's reasonableness and "his freedom to make, in appropriate circumstances, a *gratuitous* choice." Braithwaite calls this "an *absolute* freedom, an *inescapable* freedom, an *unavoidable*, an *ineluctable* freedom." (p. 43) And he goes on to argue that it is not logically possible to construct a computer "so that it would imitate Matthew's gratuitous freedom of choice." (p. 44) These audacious conclusions require further discussion.

In the first place, the difference in the two families of credal sets is significant only if one accepts Braithwaite's account of games with nature. But no independent arguments are given for this account. In particular, Postulate 5 is completely unsupported. Surely there are many other possible accounts of games with nature which would yield the same family of possible credal sets as Rule AB. On Braithwaite's account, assuming the standard example, $U_L(m_1, m_2)$ implies $C_L(m_1 \vee m_2)$ since the credal set for the former, $[\mu_1 \cup \mu_2]$, ignores μ_3. The reverse implication fails because $C_L(m_1 \vee m_2)$ is compatible with both $C_L(m_1)$ and $C_L(m_2)$. However, if one considers only Luke's *strongest* creed, then the reverse implication holds and the credal set for $C(m_1 \vee m_2)$ should be $[\mu_1 \cup \mu_2]$ rather than $[\mu_1, \mu_2, \mu_1 \cup \mu_2]$, as given by Postulate 5. This account should yield the same family of credal sets as Rule AB.

Braithwaite early on abandons a theory which has Luke selecting a *single* game from his *Ur*-game in favor of the theory of credal sets. Yet it should not be difficult to devise a set of rules for determining a unique game on the basis of Luke's strongest creed. Braithwaite surely has not shown that this cannot be done. Certainly such an account of games with nature need not exhibit significant formal differences from games with persons.

Even within the context of his own theory, Braithwaite notes that Luke's credal set is the same if Luke is merely uncertain between m_1 and m_2 in a game with nature or if he is playing against a person for whom M_3 is inadmissible while M_1 and M_2 are optimal. Thus an alternative to endowing Matthew with the power of making gratuitous choices would be simply to make Luke uncertain as to which action Matthew will perform. Braithwaite rejects this move, but for reasons which seem to have little connection with the theory of credal sets. If Matthew faces two optimal actions, so that no principle based solely on utilities could distinguish them, then, according to Braithwaite, Luke has two options. He can treat Matthew as a randomizing automaton which selects an action according to some probabilistic law; or he can treat Matthew as a person endowed with the power of making gratuitous choices. In the former case, Braithwaite claims, the uncertainty "can be resolved by guessing the method of randomization and putting the guessed chances into a new basic decomposition describing the choice situation." (p. 50) In the latter case, it is claimed no such resolution of the uncertainty is possible.

The main difficulty with this argument is that it is hard to see why guessing should be a legitimate method for resolving uncertainties about probabilities and not a legitimate method if one were to believe that Matthew's choice follows no probabilistic law at all. Indeed, since when is guessing a legitimate method of resolving any uncertainty whatsoever?

Finally, the two categories of deterministic and probabilistic are generally taken as being exclusive and exhaustive – at least with respect to physical systems. Braithwaite is really asking us to consider a third category, i.e., non-deterministic, non-probabilistic processes. There are several problems raised by the introduction of this concept in the present context. One is that no finite amount of data regarding Matthew's choices among optimal actions could refute the hypothesis that Matthew is a hierarchical randomizing automaton of some complexity or other. It does not follow, of course, that this hypothesis could not be rejected on generally empirical grounds, but the reasoning would have to be fairly complex. A second difficulty is that the connection between possible non-deterministic, non-probabilistic processes and human freedom of choice in any standard sense, has not been made. It is now generally agreed that the existence of physical indeterminism is neither necessary nor sufficient for human freedom of choice. The same sorts of arguments would seem to apply to the relation between the existence of non-deterministic, non-probabilistic processes and freedom of choice. Thus it would seem that the use of the notion of freedom of choice in any but a metaphorical sense is out of place in Braithwaite's discussion. Game theory itself provides no means of choosing among optimal actions. So relative to the game theory context, there is indeed some 'freedom of choice' remaining. But whether the selection of a single action is then to be made by a randomizer or some other process, and whether these processes are relevant to the general concept of freedom of choice are further, independent issues.

Indiana University

I. J. GOOD

COMMENTS

I found this paper most stimulating and original and would like to make a few comments and ask a few questions. These comments and questions relate to the typed version of November 1, 1969, which I was privileged to read in advance of the lecture.

(i) If the logic of the paper is intended to be consistent with a Bayesian philosophy, as I think was suggested, then there is some question of how 'certainty' is to be interpreted. Many card-carrying Bayesians consider that no empirical proposition can be certain, with the exception of some propositions that refer to private experiences, such as 'I feel pain'. Certainty will often be expressed by a Bayesian, but on closer questioning he would admit that this was only a manner of speaking used for the sake of simplicity.

(ii) There is a mention of the inconsistency of a variety of decision techniques, and one of the techniques mentioned is the principle of maximizing expected utility. I wonder whether Professor Braithwaite has in mind a neat article by John Milnor called 'Games against nature', in *Decision Processes* (ed. by R. M. Thrall, C. H. Coombs, and R. L. Davis, Wiley, 1954), pp. 49–59. This paper attacked Laplace's rule, and by this was meant a postulate of equiprobability. Such a postulate is not regarded as a rule of application of the axioms by the school of modern Bayesians to which I at any rate belong, and I wonder whether Professor Braithwaite is alleging some other inconsistency.

(iii) When a preference is evenly balanced between two actions, so that act A is not preferred to B and B is not preferred to A, then you might generally choose act A only by habit, as Professor Braithwaite said in the discussion. But this would mislead the kind of experimental psychologist who ignores all introspection. This shows that your preferences cannot be entirely determined from your non-verbal actions alone, and that your assertions need to be taken into account also. We rightly feel that in most circumstances actions speak louder than words, but it seems to be the other way round in this example. Although speech is a form of

Leach et al. (eds.), Science, Decision and Value, 67–69. *All Rights Reserved.*
Copyright © 1972 by D. Reidel Publishing Company, Dordrecht-Holland.

behaviour, we do have here a serious difficulty for anyone who wishes to adopt a purely behavioural criterion for the discovery of preferences.

(iv) I think preference between acts always involves an implicit judgment of probabilities of future events. My question here is (a) whether Professor Braithwaite agrees with this view, and (b) whether he feels that it is relevant to his thesis if he does agree.

(v) Regarding the *Angst* involved in the uncertainty about the existence of God, about survival after death, and about human destiny; I do not see how this differs from the *Angst* engendered by an absence of states of certainty. I would appreciate the distinction if the former means that the probability varies rapidly from time to time, since this would be typical of the onset of a nervous breakdown, but I understand that this is not Professor Braithwaite's meaning. I am asking this question with a theory of partially ordered probabilities at the back of my mind.

(vi) Professor Braithwaite made a suggested distinction between treating one's opponent as an automaton and as a person. He says that even if an automaton had a finite hierarchy of randomizing devices, each randomizing the parameters of the device one lower in the hierarchy, that this would still be an automaton. He seemed to me to imply that a person is different in that his gratuitous behaviour involves an infinite hierarchy or some other technique not available to an automaton. Since I believe that a person is capable of only an effectively finite number of states I find this distinction difficult to swallow. Nor do I believe that any mathematical argument could ever prove that there is a fundamental distinction between a person and an automaton. There might very well be a distinction but it seems to me to be a metaphysical one, depending on questions of consciousness, pleasure, and pain. I raised this point in the discussion and Professor Braithwaite said he had not meant to make this distinction, but then I need further clarification of the point that he was making. The basic reason for my puzzlement is that it seems to me that a sufficiently complex automaton ought to be capable of behaviour that would seem to any man to be gratuitous simply because it would be too difficult to formulate a comprehensive probabilistic model of the automaton. Perhaps this is all that Professor Braithwaite really intends to convey by a *person*: an org (organism, organization, or complex machine) that is too difficult to analyze completely.

(vii) Likewise I should like to ask whether *nature*, in discussions of games with nature, is defined thus: a complete probabilistic specification of its mode of play can be made in a comparatively simple manner, the specification being the same when the name of the game is the same. If the specification is complicated but not totally impracticable then the opponent is an automaton; and if totally impracticable then the opponent is a person, or as good as a person as far as the strategy of the game is concerned.

C. WEST CHURCHMAN

MEASUREMENT: A SYSTEMS APPROACH

(This paper is a review of an earlier work of mine, *Theory of Experimental Inference* (TEI), on its twenty-first birthday, in order to assess its promises.)

It has become popular in philosophical literature to refer to a certain formulation of the problem of induction as 'Hume's problem', with the inevitable questioning whether Hume actually had such a problem as the formulator proposes. I should characterize the central theme of TEI as a concern with 'Kant's problem', with the inevitable admission that Kant himself never faced the problem as I formulated it. 'Kant's problem' is stated exactly in the middle of the text:

... we now seem forced to the following paradoxical conclusions:
   1. We must agree with Kant that the construction of an intelligible world out of the immediacies of sensation demands a certain *a priori* equipment, *i.e.*, demands the assumption of certain principles not derivable from the elementary facts of experience.
   2. But we must disagree with Kant that *a priori* laws are known intuitively, and that their verification is independent of observation.
   Thus our story seems to have led us to what many would be inclined to call an *impasse*; we are required to assume *a priori* law to learn about the world, but we cannot determine the correct *a priori* until we have learned about the world. (TEI, pp. 144–5)

As the text indicates, this was not Kant's problem, but rather post-Kantian. Kant, with a great deal of wisdom and foresight, attempted to solve the problem of the 'correct' *a priori* by the method of necessity: the *a priori* forms and categories are necessary conditions for any meaningful experience. In schematic form, Kant's argument is that:

(1) all knowledge begins with experience,

(2) *a posteriori* knowledge exists,

(3) a certain kind of *a priori* knowledge is a necessary condition for *a posteriori* knowledge; therefore

(4) this kind of *a priori* knowledge exists.

TEI essentially attacks the third premise of this argument by weakening it to the statement:

Leach et al. (eds.), Science, Decision and Value, 70–86. *All Rights Reserved.*
Copyright © 1972 *by D. Reidel Publishing Company, Dordrecht-Holland.*

(3*) some kind of *a priori* knowledge is a necessary condition for *a posteriori* knowledge.

To arrive at Kant's problem, TEI relied heavily on a theory of measurement, first proposed in its most coherent form by E. A. Singer in his seminars, and later expounded by him in his *Experience and Reflection* (1959). Since I am far more interested here in where one goes from Kant's problem rather than how one gets there, I'll be brief in summarizing Singer's argument. Singer translates Kant's account in his *Critique of Pure Reason* (1781) of how the mind succeeds in experiencing into an account of how a measurer succeeds in measuring. A measurer must have a measuring instrument which enables him to compare certain characteristics of the observed world. But such comparisons are not possible unless the measurer assumes that his instrument is properly calibrated. The assumptions which he makes are the *relative a priori* of a specific occasion of measurement. They are relative because, in another context, the same or another measurer may check the calibration assumptions, but in order to do so he must also make some relative *a priori* assumptions. Thus *some a priori* assumptions are made in every occasion in which measurement takes place, but the exact form of these assumptions varies *depending on the measurement task*, *i.e.*, depending on what the measurer is trying to accomplish.

The very language and form of this argument reveal where one has to go from Kant's problem, though TEI, understandably lacking in references to later literature, does not delineate the pathway clearly. If the correct *a priori* assumptions depend on what the measurer is trying to do, then a measurer is just like any other decision maker in social systems: he ought to select the alternative which best serves his purposes. As system science would say, the problem is one of strategy (or tactics).

The consequence is that Kant's problem becomes a particular problem of prescriptive science, *i.e.*, that science which deals with the selection of the right alternative for the pursuit of specific goals. As TEI pointed out, this consequence is by no means astonishing news to the philosophical community, because it had long since become the basic theme of pragmatism, *e.g.*, in Dewey's instrumentalism. But it can safely be said that no pragmatist of the first part of this century, including the author of TEI and his philosophical mentor Singer, ever realized the enormous complexity and confusion that the development of a sound prescriptive science entails. In some ways, those philosophers who stopped short of Kant and elected to devote themselves to Hume's problem showed more prudence if not more valor. Apparently Hume's problem can be investigated within the domain of logic, albeit an 'extended' logic of some kind. But Kant's problem can only be investigated in the context of all the sciences. If today's 'systems science' is what yesterday's pragmatists had in mind when they suggested that truth is what 'works out', then engineering, biology, economics, behavioral science, computer science, and ethics are all involved in the enterprise of defining and implementing the 'workable'. Nor could one say, as we are about to finish off the third quarter of our century, that the rather facile promises of the midcentury system sciences[1] have come anywhere near being kept. We have learned a great deal, to be sure, but most of it is about how little we know.

Here are a few examples of the promises of the 1950's. In von Neuman and Morgenstern's *Theory of Games and Economic Behavior* (1953) it was suggested that the clear pathway to an understanding of optimal decisions in conflict situations consists of starting with the simplest 'two-body' problem, *i.e.*, constant sum two-person games, and working up to the 'many-body' problem of non-constant sum n-person games, a 'working up' that is still in its infancy without any very significant application to today's world of conflict. In mathematical programming, it was suggested by many that once computers became large enough, and certain knotty problems of uncertainty and non-linearity were solved, one could model large, complex organizations and deduce optimal strategies from the models. Today's computers are indeed large and are handling programming models with two million variables and 35000 constraint equations, but no one would dare claim that we have the capability of 'modeling' whole organizations with anything like a reasonable accuracy. As for artificial intelligence, we still must use people to translate common languages, and the chess champion of the world is remarkably human even though he is a red rather than a white piece.

TEI also made one such promise, really far more grandiose than the rather moderate ones just mentioned. In order to "get beyond" the apparent relativism of many pragmatists, wherein truth becomes the servant of special interests of persons, societies, or cultures, TEI suggested the need to develop a theory of 'ideals', of non-attainable objectives which somehow capture the essence of all men's values. Here is the promise:

... we suppose that it is possible by an examination of the histories of societies with respect to their aims and conflicts, to determine predominant purposes expressive of the aims of man, not as viewed from one age or social group, but as viewed throughout all changes or societies ...
... we are demanding an experimental science of history as a basis for determining the predominant purposes. (TEI, p. 262)

The statement is certainly characteristic of the optimism of 1950 with regard to science and its capabilities. When demands of this type are not met, even to a modest degree, the demander can do no more than challenge, as I did in 1961:

What seems to be the distinctive contribution of the twentieth century to the theory of human progress is the recognition that there can be no progress without conflict. The challenge is to develop a science which can understand what this means. (*Prediction and Optimal Decision*, p. 380)

That is, he continues to challenge until he begins to become somewhat self-conscious about his own arrogance, as I did in 1968:

Who is challenging reason? What is the justification of this challenge? I am challenging it in this book, but I surely don't know whether I am justified. (*Challenge to Reason*, p. 218)

Now that the boy has grown up, it's time his parents stopped challenging him, and began to think about his career. And in some ways it does appear as though the career is already well established as a career of 'systems science' or the 'systems approach'.[2] With its rather modest beginnings in the 1940's, systems science has now spread out to the study of many large social organizations in order to assist them in gaining their desired goals. Even though the pace has been slower than the optimism of 1950 predicted, it is nonetheless impressive, as system science moves into local, federal, and international problems of poverty, pollution, and peril. Why shouldn't the challenger still feel that his challenge will eventually be met, even though the time scale may be centuries rather than decades?

To respond to this question I need to say something about the role of philosophy in systems science. In a sense, the question of the proper role of any discipline in this area may not be the right question to ask, because one of the strongest lessons we have learned in system science is that a systemic approach to social problems is – not *inter*disciplinarian – but *anti*-disciplinarian. The disciplines by their very nature must take the social problem and shape it into their own perspective, *i.e.*, must distort it by their representation of reality. Thus engineering wants to convert the social problem into a problem of engineering change, *e.g.*, devising cybernetic feedback systems. Economics wants to convert the problem into one of maximizing economic benefit minus cost, while behavioral science tries to see the problem as one of easing tensions, and so on. In each case, there is a very high probability that the 'solution' which the scientist offers is no solution at all, just because the distortion is so great. To use Thomas Kuhn's concepts in his *The Structure of Scientific Revolutions* (1962), systems science must abhor the paradigm. But each discipline's paradigm is like an unexamined but firm policy of an organization, the very weakness which systems science tries to ferret out and eliminate.

And yet the question of the proper role of each discipline is a practical and sensible one, simply because all of us have been educated and trained along certain lines, and cannot realistically escape from the mode of representation and the paradigms of our disciplines.

With advent of symbolic logic, many philosophers began to interpret their role as one of clarification of concepts and inferences; they saw themselves engaged in explaining as completely as possible what a word or a problem really means, with the common sense notion that, unless we know clearly what we're talking about, the rest may well be nonsense. Along with this reconstructive effort came inevitably the policy of studying manageable problems, which could be isolated, so to speak, from the rest of the system.

On both counts – clarification of concepts and determination of manageable problems – the well trained philosopher faces considerable difficulties in aiding the development of system science, no matter how well he is motivated. One lesson every eager system science student quickly finds as he enters the real world is that clarification of meanings and problems comes at best at the end of the study, not the beginning. We always do our applied work in an atmosphere of confusion, and a concentrated effort at the outset to say precisely what we're trying to accomplish is inevitably premature and leads to a precise solution of precisely the wrong problem.

I am not suggesting, however, that the philosopher-logician has no role to play in systems science so long as he realizes that his is not the dominant, first-place role, and so long as he realizes that every effort to make a problem manageable threatens a loss in the effectiveness of the solution. But I am also interested in eliciting the aid of another kind of philosopher who is more often than not scorned by his more precise colleagues. This is the *critical* philosopher, where 'critical' has all the richness of meaning which Kant ascribed to it. Of course, it must be granted at the outset of this exploration that we do not know precisely what Kant meant by 'critique', or, more to the point, what we today ought to mean by it.

'To criticize' means 'to judge'. Traditionally, 'judgment', as well as 'decision', meant to reach a conclusion in the context of a dispute between two protagonists. The conversion of the tradition of a three-party judgment (judge and two disputants) into a single reasoning mind occurs in Kant's *Transcendental Dialectic* of the first *Critique*. Here Kant himself designs both the opposing theses and their respective 'defenses'; furthermore, he designs the judgment which he claims resolves the conflict of ideas. Reason, therefore, plays the role of resolving the dispute which it itself has created.

Kant's step is a very important one relative to the use of criticism in system science. If philosophical criticism has a role to play, Kant's invention suggests that criticism is to be designed as an explicit methodology, as self-contained as the models which the system scientists use. And, like the models, it runs the extreme risk of being totally out of touch with the rest of social reality, simply because it creates its own dispute and resolves its own problem.

A specific illustration may help to understand the role of criticism in system science, a very mundane, 'practical' example. Suppose one wants to aid the purchasing department of an industrial firm in developing an optimal inventory control of its supplies. We assume at the outset that 'optimal' is solely dependent for its meaning on the concept of profit maximization of the firm. The manner in which systems science should proceed is taught to every schoolboy in operations research as one of his first lessons. He must 'gather some data' which tell him the pattern of demand for an item over time, the cost of holding an item in inventory, the cost of a shortage, and possibly answers to other questions. But a little reflection (often lacking in practice!) reveals that historical demand may not be relevant, because that part of the system which uses the items in inventory may be poorly designed. What's the point of servicing a badly run component of the system? This reflection apparently leads to the awkward conclusion that the demand data cannot be collected accurately until the system scientist has optimally designed the component which generates the demands. Or, consider the cost of holding inventory. A part of this cost is called the 'cost of capital', which means the loss in gross returns incurred by tying up funds in inventory. It (like all costs in systems science) is a lost opportunity cost. To measure it, one must determine how the firm might have used the funds had they not been spent on inventory. (For those concerned with the paradoxes of 'counter-factuals', it may be interesting to note that the cost of holding inventory is only determinable by establishing the validity of a counter-factual). But the epistemological problem of measuring this opportunity cost should more accurately be phrased as one of determining how the firm *ought* to use its available funds, for the financial managers may be quite negligent in this regard. Thus, all the system scientist has to do to solve the purchasing department's problem is to solve the problem of how the firm should use its available capital, *i.e.*, how the firm should be run!

Such a conclusion must appear altogether satisfactory to those positivists who told us young philosophers of the 1940's that we could never determine the 'ought' from the 'is'. There is simply no way of using direct observation to determine any of the data which system science uses. But TEI argued that *all* science is in a position exactly like that of system science, since the conduct of science is a system science problem. But here I am less interested in the generalization than I am in how system science might go about resolving the apparent paradox.

TEI's answer is the one commonly held by many philosophers of science who wrote before and after its publication:

... the time has come to recognize the circularity, or spiral form, of science. It is perfectly proper to consider one phase of nature as though it were known, while we develop another phase, as long as we do not make this a permanent state of affairs. (TEI, p. 216)

Thus it is 'perfectly proper' to assume that certain managers know how to run their piece of the business, while we develop an 'optimal' inventory policy, so long as we don't make the assumption permanently, *i.e.*, so long as later on we get a chance to study their problems. When this chance occurs, we may very well come back and revise our earlier estimate of the optimal. This idea is the basis on which Singer describes all 'data' as being in the optative mood: "... let us for the time being take the cost as lying in such-and-such a range."[3]

The role of criticism now becomes clearer. Its aim is to challenge the assumptions of the systems science measurer at the appropriate time. The role of philosophical criticism is one of asking for the criteria of the appropriateness of a criticism and its timeliness. The aspect of this question which interests me most at the present time is the question whether the whole process of inquiry 'gets anywhere', *i.e.*, exhibits the mark of progress, even if all assumptions are 'systematically' questioned. It is interesting to note how quickly TEI glides over this question. In the quotation just given, there are the phrases 'circularity, or spiral form...' But the 'spiral form' is only one manifestation of circularity. Spirals 'get somewhere', though not necessarily with the 'upwards' motion that TEI obviously assumed.

The confidence that system science is an upward moving spiral is based on extremely questionable evidence. One may point to the 'success' of physical science in reducing, say, the probable error of the velocity of light *in vacuo* from several thousands of miles per hour to less than one mile per hour in the span of two centuries. But has science 'progressed' in this period of time? To say that it has is somewhat like arguing that a firm has become more successful because its product has become purer, when in fact no one wants the product in the first place. The awesome question is whether science, in its pursuit of more precision, has not totally forgotten what business it's in, and especially forgotten to concern itself about its own survival. To reduce probable errors while producing nuclear power is not necessarily to progress in the system science sense. Indeed, the entire last century of science might be characterized by what the system scientist calls 'suboptimization', *i.e.*, by the effort to become much better in one component, using that component's assigned scoring system, but as a consequence becoming much worse in the whole system according to the whole system's scoring system.

So the question of philosophical criticism becomes a classical one: what are the qualities of the whole system which guarantee the survival and the progress of inquiry? The ghosts of Descartes and Leibniz, together with their quaint cosmological and ontological proofs of the guarantor, have come back to haunt the system scientist. As far as I can tell, no system scientist today can adequately respond to Descartes' critical thesis that the 'guarantor problem' is a central issue for all system science which assumes that it is 'getting somewhere' in solving problems of pollution, transportation, education, etc. Having long since abandoned hope of a simple proof of God's existence, we rather foolishly abandoned the whole question, a tactic which we would never think of employing for other serious questions. If God is that particular combination of characteristics of human society which guarantees progress and survival, then what is God and does He exist? Or, how can God be designed, and what is an optimal God?

Now one characteristic of philosophical criticism is that it can always be turned on itself. My more precise friends, who are not always precisely my friends, may be disappointed but not surprised that in approaching old age I've begun to talk theology. But that is not the last word of approaching senility. Not only has systems science neglected to guard itself against a system destroyer, but it has also neglected to guard itself against systemic immorality.

To understand this point, it is necessary to represent system science, not as an abstract set of models, computer programs, or flow charts, but as a behavioral set of relationships. Briefly, the *system scientist* uses conceptual behavior in order to influence the actions of a *decision maker* in order to maximize the benefits of the system for the legitimate *clients*. The relationship between the three parties of this arrangement (system scientist, decision maker, and client) is a relationship which is suspect on moral grounds. The system scientist calculates that configuration of people and physical resources which will best serve the client, and then attempts to influence the decision maker to change people and resources from their present state to an 'optimal' state. To accomplish this aim, he looks on people, as well as resources, as means to the accomplishment of social goals. In true Benthamite style, he realizes that some clients' values will have to be sacrificed in order to gain the most the system can accomplish for all legitimate users. Specifically, the system scientist's behavior is one of using at least some people as means only, not as ends.[4]

Now we return again to Kant and another of his problems, the nature of the Good Will. Kant argues in the *Fundamental Principles of the Metaphysics of Morals* that a form of the moral law is: "So act as to treat humanity, whether in thine own person or in that of another, in every case as an end withal, never as a means only."

In order to keep this part of our story brief, suppose we dub Kant's moral law as L_K. Consider now the following policy statement of the system scientist, the so-called trade-off principle: "In order to estimate an optimal decision, it is necessary to place weights on various outcomes and to choose that alternative which maximizes the weighted outcome values, at the inevitable sacrifice of less important outcomes." Call this principle L_S. The unexamined thesis of system science is that L_K is true, and that L_S implies not-L_K. If the thesis is true, systems science is essentially immoral. Evidently, much needs to be done to explain this thesis, and here we should indeed enlist the aid of the formal philosopher, who might help us understand better the concepts inherent in L_K and L_S, for there is no reason which I can see why L_K should not be made as formally precise as L_S has become.[5]

Thus the older man, quite naturally, finds deep and serious doubts about the younger man's promises and ambitions. But he has by no means given in to the younger's philosophical enemies, those who, through pure negative feeling, spouted nonsense about the limitations of science, the 'essential' elusiveness of metaphysical inquiry, and so on. If anything, the older man has become even more serious than the younger about the importance of science.

It is this very seriousness which I want to examine in the closing part of the essay, for it too is badly in need of philosophical criticism, and it too has its own antithesis. The opposition I am referring to has been so well expressed in literature that I'll begin there. In a recent article, James Hillman paints the opposition in vivid colors.[6] "The senex archetype", he says, "is the wise old man, concerned with the development of the human race, a Roman Saturn, not ambivalent but definitely good". His example today might well be the serious system scientist in HUD or HEW who, joylessly but with utter dedication, seeks to relieve the lot of the poor in the ghettos. But as Hillman points out, the senex also had its negative side: "... he feeds himself insatiably from the bounty of his own paternalism." Here is the vivid description:

His *moral aspects* are two-sided. He presides over honesty in speech – and deceit; over secrets, silence – and loquaciousness and slander; over loyalty and friendship – and selfishness, cruelty, cunning, thievery and murder. He makes both honest reckoning and fraud. He is god of manure, privies, dirty linen, bad wind, and is also cleanser of souls. His *intellectual qualities* include the inspired genius of the brooding melancholic, creativity through contemplation, deliberation in the exact sciences and mathematics, as well as the highest occult secrets such as angelology, theology, and prophetic furor. He is the aged Indian on the elephant, the wise old man and 'creator of wise men', as Augustine called him in the first systematic polemic against this senex archetype. (Hillman, p. 319)

This passage can be regarded as a dramatic version of my earlier charge that the serious systems scientist is at once both beneficent and immoral.

The 'puer' archetype, on the other hand, is characterized by freedom, play, adventuresome heroism, and, above all, *no* development or progress. The vitality of the puer lies in the game itself, and not what games may lead to in terms of a person's health. Today's flower people and street people are a clear exemplification of the puer as well as its negative aspect; there is the threat that with an 'only-puer' concern 'the world as world is itself in danger of dissolution into the other-wordly'.

There is also Joyce's version of the same opposition of archetypes in the fable of the 'Ondt and the Gracehoper':

The Gracehoper was always jigging ajog, hoppy on akkant of his joyicity,..., or, if not, he was always making ungraceful overtures to Floh and Luse and Bienie and Vespatilla to play pupa-pupa and pulicy-lulicy and langtennas and pushpygyddyum and to commence insects with him...

while:

The Ondt was a weltall fellow, raumybult and abelboodied, bynear saw altitudinous wee a schelling in kopfers. He was sair sair sullemn and chairmanlooking when he was not making spaces in his psyche, but, laus! when he wore making spaces on his ikey, he ware mouche mothst secred and muravyingly wisechairmanlooking.[7]

The relevance of the literary contrast to this essay's conclusions are all too apparent. Is 'science' a senex or puer, an Ondt or a Gracehoper? Is 'science' a serious matter? In both TEI and this essay, I have assumed that science is obliged to look after its own interests, and have inferred that this obligation sends it inevitably into the dangerous quagmires of systems science. But there is another 'model' of science, Joyce's model, perhaps, which says that the serious side of life owes it to the non-serious to sustain its life of song and joy, because

> Your feats end enormous, your volumes immense,
> (May the Graces I hoped for sing your Ondtship song sense!
> Your genus its worldwide, your spacest sublime!
> But, Holy Saltmartin, why can't you beat time?
> (*Ibid.*, p. 419)

It would be very much in the spirit of TEI to attempt to formalize the opposition which literature has so artfully expressed. But first of all, some background remarks regarding such a formal schema are essential. TEI borrowed directly from E. A. Singer's idea that it is possible to use a formal classification of responses to a philosophical question in order to clarify what the history of philosophy has to say on the matter and to make maximum use of its arguments. The point that Singer was making is that philosophy is essentially a reworking of its own historical disputes, and that a careful definition of a dispute may be one of the most powerful methods of philosophical work. It was not Singer's (or my) intention to put philosophers into pigeonholes, nor to suggest a classification of their intellectual lives.[8] We were not attempting to prove that Leibniz was a 'rationalist', for such proof would have involved a deep and scholarly examination of the Leibnizian corpus and its commentary. Rather, we developed a fairly precise set of answers to the classical problem of *a priori* knowledge, one of which we called 'rationalism', with some textual evidence to justify the label. We then thought we could find cogent arguments in Leibniz's writings which defended the rationalist answer. Our 'reworking' of these arguments consisted in a critical examination of their content, where 'critical' has the meaning already described in this essay.

One other remark is highly important for understanding the philosophical approach. TEI assumes a realist rather than a nominalist position with respect to the definitions which are the heart of the formal classification. The particular form this realism takes is the assumption that definitions are in part based on a tradition with an intellectual content; a proposed definition, therefore, is rarely arbitrary, and may be judged by criticism to be 'right' or 'wrong', depending on its subtle relationship to the relevant traditional content. Like all responses to scientific questions, no definition is ever completely 'right', and subsequent history will keep modifying and (hopefully) improving the philosophical dictionary. I don't intend to defend the realist position here, except to mention that the argument in its defense is pragmatic: wrong is the man who runs down a hotel hallway yelling "Fire!" and later, in Humpty Dumpty style, declares that when he says "Fire!" he means, "It's raining outside". But the realist position has a very important consequence which I think TEI and other related work[9] of that time failed to emphasize enough: the reliability and usefulness of a formal definition depends very strongly on our knowledge of the phenomena to which it refers. If at a stage of history or culture, men are largely ignorant about a certain aspect of nature, then formal definitions are apt not only to be useless but dangerously deceptive. A good case in point is the economist's definition of 'preference', which has become a foundation of a great deal of work in decision theory.[10] Our knowledge of the psychological phenomena of human preferences ('values') is still very weak: e.g., we have only a very vague notion as to why men prefer deadly conflict to peaceful settlement. Hence, formal defining may be worsening rather than bettering our total comprehension. This may be why formalizers have tended to shy away from defining 'love', 'happiness', 'grace', and other elusives. However, I am not asserting that formalizers should stay away from these elusive areas; but their effort should be one of revealing that we do not know, rather than assuming that their definitions are precisely right or arbitrary conventions.

Back now to the literary opposition. The two positions discussed there are describable as the serious-solemn vs. the joyous-jocular. Etymology's contribution tells us that 'serious' comes from the Latin serius meaning 'heavy', whence 'weighty', and a later meaning, 'not to be disregarded'. 'Solemn' is from the words 'sollus', meaning 'whole', and 'annus', meaning 'year', whence 'performed with due formalities'. For brevity's sake, I'll assume that there is etymological justification for the hypothesis that the pair 'serious-solemn' describes the mind which attempts in a formal (*i.e.*, systematic) manner to take into account all the relevant consequences of human decisions. In other words, a serious-solemn mind is the ideal systems scientist which I have been discussing in this paper. It should be noted that the adjectival pair describe an intention and not an actual deed: the serious-solemn mind intends to the best of its ability to approach life's decisions from a system's point of view. Thus we can regard the past twenty years of extensive research in systems science as providing a far richer meaning to the concepts of seriousness and solemnity than, say, was available to a father of systems science, Jeremy Bentham.[11]

When we turn to the other pair, the joyous-jocular, the matter is quite different. Our knowledge of the joyous mind, I think, is extremely limited. Some would like to see it remain so, because of a modern version of the classical hedonist paradox. Imagine, for example, the 'breakthrough' which would produce a joyous computer! It would be the calamity which would rob us of the joy of human living. Scientists, of course, have been sniffing around the outward skin of the joyous-jocular by trying to study creativity, or to define humor, or occasionally and badly to define 'pure research'. To say, for example, that 'pure research' is the 'satisfaction of intellectual curiosity' is sheer nonsense, because it leaves out all the immense and absorbing joy of the discovery and sharing of a new idea, a joy that is deeply human. To illustrate what I mean, one might try to define 'trivial' in a non-trivial way; after all, the joy of discovery is essentially the finding of a non-trivial aspect of nature.

But, though for the time being the meaning of joyous-jocular must remain elusive, some of its characteristics, as seen by a serious-solemn mind, are relatively clear. For one thing, the serious and the joyous share one aspect very strongly: a reliance on a thoroughly trained mind. *Magister Ludi*[12] and *Zen and the Art of Archery* are examples on the joyous side of the comparable preparations of a mind which seriously sets out to improve society.

The deep chasm between the two pairs can also be partially explained in serious terms: the serious mind views the improvement of human society as a teleological activity, a determination of the real goals and the ways to pursue them. But the joyous is ateleological; the so-called means-end framework is irrelevant to the feelings of joy and humor. Now of course the serious mind has a way of accounting for ateleology. He can represent human values in terms of man's utility for goals. In some cases, a man may place a high utility on tomorrow's goal, and no utility on any of the consequences beyond tomorrow. As we shorten the time span between the act and the consequence under these conditions, the limit is the preference for the act itself, which the serious mind might then want to call an 'ateleological' preference.

I suspect, though, that this bit of serious trickery may miss the mark, and for much the same reason that the teleologist cannot at the present time adequately handle the problem of morality. In the case of morality, he would like to say that as systems become improved, we will gradually reduce the risk of treating men as means only. But a strong moralist would object that no amount of eventual 'do-gooding' will ever remove one bit of the evil that was perpetrated in getting there. So the joyous-jocular mind is likely to find that the serious mind's account of ateleology is a bit of a joke. For a Western trained philosopher who has seriously learned his teleology through pragmatism, modern biology, psychology, sociology, and economics, it may be entirely too difficult to reverse the path and get back to understanding ateleology.

But none of the elusiveness of the joyous prevents the serious from undertaking its usual classification, which I'll do as an allegro penultimate movement of this paper, in a solemn-jocular mood. Let us say that science is non-trivial explanatory behavior, and that the philosophy of science is non-trivial explanations of science. (This may sweep in more rubbish than many philosophers might wish, but they can define it out the door again if they wish.) Our serious mind immediately sees four philosophies of science:

(1) One can try to find joyous explanations of joyous science. To be sure, many philosophers tend to be dead-pan in their joyousness, but it would be safe to say that those who spend their lives in trying to resolve the paradoxes of physics and logic provide the arguments for justifying this kind of philosophy.

(2) One can try to find joyous explanations of serious science. Here a good example is the work now being done in the utility of information and Marschak-Radner 'Theory of Teams'. The workers apparently have no serious intent that their results will find a use. In some ways, this paper tends to fall into this category.

(3) One can try to find serious explanations of joyous science. Professional philosophers have not tended to interest themselves in this area, but those who find themselves in the Washington scene these days do discover its extreme importance. For many joyous scientists find that Congress is reluctant to support them as it used to do; the awkward and very serious question is which projects should die, and what fields are less important than others. It is to be noted that joyous scientists all tend to scorn joyous efforts in other disciplines, but few of them ever meant their scorn to be taken seriously.

(4) One can try to find serious explanations of serious science, *i.e.*, to adopt a 'system approach' to science as a social institution. This is happening, will happen more, and may be a very dangerous trend unless the serious mind realizes that it has left out completely its very alive deadly enemy, the joyous.

In conclusion, I'd like to pose the critical problem of the serious philosopher studying serious science in another way, mainly because the problem is still partially unstructured and needs to be discussed from different points of view. Empirical science can be said to have used three kinds of basic models in its description of nature: the deterministic, the random (uncertain),[13] and the teleological. At one time, *e.g.*, a century ago, there was a conviction that the last two would eventually be reduced to the first, *i.e.*, that randomness and purpose are pseudo-phenomena, eventually to be removed by more precise measurements of nature. One of the twentieth century's major revolutions, I think, was to conclude that none of the three is basic, and that the choice of which one to use in a given context is a strategic choice.[14] There is no doubt of the richness inherent in these basic models, and the fact that one can pass freely from one to another has immensely increased this richness. But there is still the largely undefended metaphysical thesis that all significant aspects of nature can be described within the framework of the three models. The antithesis might state that there is a fourth 'box',[15] which is essential in the understanding of the natural world, but which science has never opened. The list of natural phenomena which might be hidden in this box is very large. I've already mentioned morality and joy in this paper. Perhaps one reason why Carl Gustav Jung's psychology has been neglected by so many psychologists is that it cannot be stated in a teleological language; *e.g.*, the very central concept of the anima and the feminine principle may be essentially elusive if one restricts oneself to teleological terms. As another example, universities are rather dangerously getting into short and long range planning, without any adequate understanding of what 'education' means. So long as they restrict themselves to the notion that education means information transmittal, then they can play it safe, but there are very loud young voices, sometimes reduced to tears by gas, who deny that information transmittal is the essence of education for human living.

And finally, and strangely, there is 'inquiry' itself. Seriously to set out to design an inquiring system may lead one to conclude that some of the essential features of this design cannot be expressed by the concepts of determinism-randomness-purpose. But far from concluding that this ends the matter, or that some aspects of the creativity of pure science are essentially elusive, I'd rather say that the young man's career is laid out for him, and that there is much hope that he'll succeed.

University of California, Berkeley

NOTES

[1] E.g., operations research, management science, utility theory, game theory, mathematical programming, and artificial intelligence.

[2] In this paper, I take 'systems science' to be the effort to improve *social* systems. In many instances the term refers to efforts to design various 'hardware' configurations (computers, space birds, etc.), but I am altogether concerned here with 'people' systems.

[3] It is not altogether clear whether the basic language of Singer's representation of science is a deontic logic. It is clear that no sentence of science, for Singer, is in the indicative mood (hence there is no problem of passing from the 'is' to the 'ought'). Indeed, in general, no one has yet worked out the basic logic of the systems sciences.

[4] The language of system science literature is replete with illustrations.

[5] Some preliminary work along the line of developing optimal strategies without tradeoffs has been done by Professor Horst Rittel at the University of California, Berkeley.

[6] James Hillman, '*Senex* and *Puer*: An Aspect of the Historical and Psychological Present', *Eranos-Jahrbuch*, XXXVI (1967), 301–57.

[7] James Joyce, *Finnegans Wake*, 1939, pp. 414–6.

[8] In defense of our critics who thought that we actually were playing the glib game of pigeon-holing, neither of us explained the method adequately, especially in the intellectual environment where there is a great deal of irresponsible labeling of philosophical positions.

[9] Particularly, R. L. Ackoff and C. W. Churchman, *Psychologistics*, University of Pennsylvania (mimeographed), 1946, which is an attempt to define a whole set of basic concepts of psychology and sociology in a decision-making framework.

[10] See, for example, D. Luce and H. Raiffa, *Games and Decisions*, John Wiley & Sons, Inc., New York, 1957.

[11] A fairly detailed definition of 'system' and specifically an 'inquiring system' is contained in a book of mine in preparation, *The Design of Inquiring Systems*, New York, Basic Books, 1971.

[12] I wouldn't suggest that the relevance of *Magister Ludi* allows one to say that science is a game. This way of characterizing science seems to me to trivialize the joyous, because 'game' has become a fairly ambigous term often connoting mere relaxation. The 'bead game' is hardly like most games we humans play in our evenings or Saturday afternoons any more than joyous science is.

[13] In recent years there has been a great deal of discussion about uncertainty models, mainly centering about the generality of the 'classical' definition of randomness, but this discussion is not of direct concern here.

[14] Among the leaders of this revolution one can mention Heisenberg, Bohr, Wiener, and Singer. The last two, working independently, showed how so-called mechanical behavior could be described in a teleological language.

[15] Obviously, the box may contain many sub-boxes.

ISAAC LEVI

COMMENTS

According to Neurath and Quine, our condition resembles that of sailors rebuilding a ship on the open sea. Ship-at-sea epistemology did not, of course, originate with Neurath and Quine. The view of human knowledge, if not the metaphor, was endorsed by the great pragmatists Peirce and Dewey. Clear vestiges of ship-at-sea epistemology are also visible in the views of Professor Churchman so that some justification can be offered for responding to his remarks within that framework, a framework which I, in any event, find congenial.

Ship-at-sea theorists tell us two things about the assumptions we employ in reaching conclusions and in forming policies:

The assumptions we employ do not, at the moment when we employ them, stand in need of justification provided that, at that moment, no serious grounds exist for questioning their truth. We have a right to be certain of the truth of such assumptions, be they reports of past observations, general laws, theoretical statements, truths of logic or truths of mathematics. We can regard acting 'as if' they are true as being without risk. Evidential assumptions are legitimately assignable maximum probability. According to the ship-at-sea metaphor, evidence corresponds to those portions of the ship which, at the moment, are not being repaired, which remain intact and which keep the ship afloat.

The second point emphasized by ship-at-sea theorists is that even though such nonproblematic, evidential assumptions can be regarded as maximally certain, their truth can by no means be considered incorrigibly secured.

Granted these two points, the central challenge to ship-at-sea epistemology is to provide a systematic account of rational ship repair and expansion. If we often do have a right to be certain of the truth of statements without incorrigible security against error, we must have some way, at least, of correcting mistakes in our evidential assumptions. There is nothing wrong in counting corrigible knowledge as knowledge bearing probability 1 provided we can fight back against the cartesian malevolent demon who would entrap us in incorrigible error. Hence, the need for a method of rational ship repair.

An account of the conduct of inquiries engaged in the correction and revision of evidential assumptions must face two interdependent problems: (1) To specify conditions when erstwhile nonproblematic 'evidence' is legitimately rendered problematic and must be removed from evidence. (2) To give an account of the rational conduct of inquiries aiming at the conversion of erstwhile problematic statements into nonproblematic evidence.

The most important contribution of the two great pragmatists Peirce and Dewey was to recognize the importance of these two questions and to attempt to reorganize the examination of epistemological investigations so as to provide for their solution. They were pioneers in this endeavor and like all pioneers are to be revered for their program rather than for the actual details of their own proposed solutions.

Indeed, certain views widely associated with pragmatism may, I think, have contributed both to the generation of difficulties for the pragmatists and to the eclipse of their influence after the invasion of America by positivism and analytic philosophy. The instrumentalism of Dewey and the apparent voluntarism of James, which are such easy targets for criticism, have by an unfortunate guilt by association led a great many to neglect the important epistemological program upon which the pragmatists embarked. Thanks to Quine, the ship-at-sea image of human knowledge has once again become respectable. Although Quine himself has not paid appropriate attention to the problem of ship repair, it is surely clear that such attention is needed. What is more to the point, it is now possible, I think, to develop an address to matters of ship repair which avoids the dubious instrumentalism so often associated with Dewey's view.

Unfortunately, Professor Churchman has gone down the instrumentalist primrose path. At least I think he has; for I have never been quite confident that I have ever captured the doctrine of the Theory of Experimental Inference by the tail. Nor am I sure that I understand the full force of his argument in his paper. The best I can do under the circumstances is to indicate points on the path where I think I have spotted him and let him tell me whether it is a case of mistaken identity.

Churchman is officially talking about measurement although his points are intended to carry much more epistemological weight than a specialized discussion of measurement would be expected to bear. In any event, his explicit references to measurement appear briefly at the beginning of his remarks only to disappear out of sight later on.

Churchman says that all measurement involves 'a priori' assumptions. His point is, of course, that to read data we require principles and assumptions which provide for their interpretation. His contention is that the correct assumption to use depend 'on the measurement task' or 'on what the measurer is trying to accomplish.'

Churchman then goes on to say the following:

> If the correct a priori assumptions depend upon what the measurer is trying to do, then a measurer is just like any other decision maker in social systems: he ought to select the alternative which best serves his purposes.
>
> The consequence is that Kant's problem becomes a particular problem of prescriptive science, i.e., that science which deals with the selection of the right alternative for the pursuit of specific goals.

This instrumentalist argument is quite remarkable. It goes substantially farther than Dewey in the Quest for Certainty when he observes:

> Objects of previous knowledge supply working hypotheses for new situations; they are the source of suggestion of new operations; they direct inquiry.

For Dewey, background knowledge is instrumental in inquiry in the sense that such knowledge is used to control inquiry. Dewey did not say, to my knowledge, what Churchman says – to wit, that because warranted assertions, previous knowledge or evidence has a definite kind of function in inquiry in general and in measurement in particular that such background assumptions are therefore adopted in a manner which best promotes the aims of the inquirer.

Ship-at-sea theorists are willing to tolerate certain types of circularity. Nonetheless, they do recognize some forms of circularity as vicious and the sort implied by Churchman's remarks as vicious to the extreme.

When a person engages in deliberation aimed at promoting some goal, the identification of options and the selection of optimal policy cannot intelligibly or intelligently be made save relative to a background of assumptions which are, for the moment, taken as nonproblematic. Churchman himself acknowledges this later on in the paper. Yet, in the passages just cited, Churchman seems to be saying that the very background assumptions themselves must be chosen relative to the very same objectives. How can this be? Either the optimality of the background assumptions would itself have to be investigated against a deeper background ad infinitum or vicious circularity threatens.

I suspect that I have totally missed Churchman's point here since I have not seen this argument reproduced in TEI and it seems clear in the later portions of his remarks that he is quite aware of the threatened regress or circularity. Yet, it is important that he be asked to clarify his intent, for on it and on it alone rests his excuse for his lengthy discussion of the past performance and future promise of 'systems science'.

But perhaps we can find a better excuse for discussing systems science (whatever that is) in the context of what started out to be a discourse on epistemology than the one offered by Churchman.

Remember that the 'a priori' assumption required for measurement or for the resolving of other problems are by no means always 'given' in the sense that they are clearly nonproblematic at the time when the problem arises and inquiry begins. A decision maker concerned to identify an optimal policy to promote his ends will often have to engage in sub-inquiries aimed at adding to the ship of knowledge background assumptions sufficiently rich to enable him to identify optimal policies. His choice of assumptions to add to his background on the basis of which he will decide on a policy will often determine the policy which he will adopt.

In this respect, how we rebuild and repair the ship of knowledge does have an important effect on how we form policies to promote given ends. Relative to that portion of the background which is already settled one might conceivably be able to regard the adoption of new items of evidence as part of a strategy designed to promote the given ends-in-view. Hence, criteria for admitting new items into the background assumptions could be viewed as a byproduct of a general theory of rational goal attainment.

Until I read his paper, it was some such view as this that I thought Professor Churchman was advocating in TEI. It captures part of the intent, I think, of a position advocated by Richard Rudner and, more recently, by James Leach. Although I believe this view to be untenable, the difficulties involved are subtle and do merit serious attention.

(1) We should remind ourselves that sometimes our aim in adding new information to the ship of knowledge is to enlarge our understanding and to gratify our curiosity. Churchman seems to think of such enterprises in the joyous-jocular vein. But understanding is surely as serious-solemn as

profit maximization, brotherly love, peace, and a chicken in every pot. There is no trickery in this bit of seriousness. It is jocular to think otherwise.

(2) Let us, however, waive this obvious point. As was observed previously, there is another important motive for adding new assumptions to the evidence. Sometimes such evidence is desired because it will improve our basis for identifying optimal policies relative to our objectives. That is to say, the background information we already have is insufficiently strong to enable us to make satisfactory determinations.

Now if we regard the adding of new evidence as part of the strategy to gain our objectives, we will not be doing what we set out to do. The addition of the new evidence will not provide an additional basis for policy formation. It will be part of the policy formed. The basis for policy formation will remain the old evidence with which we began and which we regarded as inadequate.

(3) To this observation, the rejoinder might be made that even though the new assumptions do not enrich the resources for inquiry usable in reaching a decision in the context in which it is added, new problems might arise and it might be useful as a resource in those new contexts.

But why should we regard items we adopted in an effort to promote one goal as a useful basis for promoting another? What might be a desirable addition to our beliefs for one purpose might be quite undesirable relative to other goals. It would appear, therefore, that we could never add new beliefs via inference which would then be usable as assumptions in new inquiries.

(4) Churchman has registered sensitivity to this last point both in TEI and in his paper. He suggests that in order to escape 'relativism' wherein truth becomes the servant of special interests of persons, societies or cultures, we should develop a theory of 'ideals' 'of nonattainable objectives which somehow capture the essence of all men's values.'

It is typical of relativists despairing of their relativism to turn to a quest for 'eternal verities' invariant in the beliefs of all mankind. Churchman's quest does not contribute to the solution of the difficulties in his position.

Let us suppose that we could identify such unattainable ideals. I am of two minds about the point of trying to attain the unattainable – but let that pass. As I understand it, we would be enjoined to regard the adding of new assumptions as part of a large scale strategy to promote efforts to at least approach the ideal. But, as was noted before, if the adoption of new assumptions is part of the strategy, such assumptions cannot function as part of the background assumptions which are employed in fixing upon such a strategy. But that is the purpose for which we attempt to add new assumptions in the first place.

In spite of these reservations, I do agree that a systematic account of rational ship repair should be based on a general theory of rational goal attainment. Where I disagree with Churchman, Rudner, and Leach is in my conception of the goals to be attained. Any effort to add new evidential assumptions is concerned to supply the demands for information required by a given question. More accurately, the aim is not merely to supply relevant information but information which is true as well. The point of raising such questions themselves might be to provide a basis for action aimed at promoting moral, economic, political or other practical goals. But there is a distinction to be made between the objectives of cognitive inquiries concerned with supplying demands for true information to be used as resources in future inquiries and deliberations aimed at forming policies for realizing practical goals.

On this view, both cognitive inquiries and practical deliberations are goal oriented. Hence, a general theory of rational goal attainment is of central importance to a systematic account of both the rational conduct of inquiries and the rational conduct of deliberations. Moreover, the results of cognitive inquiry are relevant to practical deliberation. Adam Smith believed that through the operation of the 'invisible hand' a society of acquisitive individuals could promote the general welfare even though the actions of the individual members of society were appraised not in terms of how well they promoted the welfare of all but only their own self interest. We all, I take it, have grave doubts about Smith's invisible hand. But there is nothing absurd in supposing that the results of one kind of activity aiming at one kind of goal might then be useful in promoting other objectives. Roughly speaking, this is true of the products of scientific inquiry which, on the view I am advocating, has its own autonomous concerns adding true information to the ship of knowledge.

In what sense of 'truth' is truth a desideratum in inquiries aimed at adding new assumptions to our knowledge. With Peirce and Quine and unlike Dewey, I would endorse what roughly passes for a correspondence theory or, more accurately, a Tarski-like theory. At any given time, those nonproblematic assumptions which are the evidential basis for cognitive inquiry looking for additions to the evidence can be exploited to provide a Tarski-like definition of truth for potential answers to the question under consideration. In choosing between potential answers, we regard true answers in this sense as preferable to false ones.

Clearly in this sense, our conception of truth changes bit by bit with alterations in our evidential and nonproblematic corpus. Is this 'relativistic'. I can do no better than to remind you of a passage from Quine's *Word and Object*:

> Have we now so far lowered our sights as to settle for a relativistic doctrine of truth – rating each theory as true for that theory, and brooking no higher criticism? Not so. The saving consideration is that we continue to take seriously our own particula-aggregate science, our own particular world theory or loose total fabric of quasir theories, whatever it may be. Unlike Descartes, we own and use our beliefs of the moment, even in the midst of philosophizing, until by what is vaguely called scientific method we change them here and there for the better. Within our own total evolving doctrine, we can judge truth as earnestly and absolutely as can be; subject to correction, but that goes without saying. (pp. 24–5)

Thus, the goals of scientific inquiry are quite as attainable as the objectives of practical deliberation – i.e., they are no less and no more attainable. There is no need to introduce syncretistic eternal verities in order to save ourselves from an objectionable relativism.

Which brings me to my final point. I do not mean to deny that fixing beliefs in accordance with the objectives of cognitive inquiry lacks moral, political, economic or practical consequences. From the time of the scandals in the school of Pythagoras to the anxieties over research pointing to the feasibility of the atomic bomb, there have been many occasions upon which the values of science have come into conflict with other values. This I take as evidence that science does have distinctive goals. Otherwise there would be no conflict save between various moral, political, economic or other practical goals.

I do not have any decision procedure for determining how such conflicts ought to be resolved. Although it is obviously desirable to have some systematic account of the rational resolution of value conflict, any adequate account will have to recognize that in each situation of conflict many highly specific parameters will have to be taken into account before a decisive solution can be offered to complete detail. This might lead to despair regarding the feasibility of decisive resolution and a certain toleration for apparently wrong decisions made by men in such conflict.

Nonetheless, I am reasonably convinced that when we face situations of value conflict, it is quite helpful to recognize the values which are in conflict. Perhaps, we should sometimes fix our beliefs so as best to promote mental health, moral uprightness and social tranquility even though the demands for true information would dictate another course. But if we do so, we should, I think, be sensitive to the fact that we are sacrificing our commitment to truth and understanding upon the alter of our moral predilections.

Columbia University

RONALD GIERE

COMMENTS

To one who had recently read *Theory of Experimental Inference* (TEI) and has considerable interest in statistical inference and scientific method generally, it came as a disappointment that the first three chapters of this book received no mention whatsoever in Churchman's review after 21 years. In these early chapters, Churchman had examined the possibility that the then still fairly new statistical theory of Neyman and Pearson provided a general basis for all inductive reasoning. He concluded, correctly, that on this view every inductive inference rests on empirical presuppositions that could not all be tested by the same methods without making further presuppositions, and so on indefinitely. The rest of the book is really an examination of possible ways of dealing with this result. From the standpoint of the last 21 years of philosophical discussion about inductive reasoning, it is unfortunate that Churchman's subsequent work follows certain trends from the latter part of TEI rather than the trend of the first three chapters. Since no one else developed this viewpoint, few philosophers today know much about the statistical methods which became standard throughout the scientific world. But perhaps the loss to philosophy has been offset by the gain to systems science, which is where Churchman is today.

Throughout his paper, Churchman contrasts the professional philosopher or philosopher of science with the systems scientist. He characterizes the philosopher as being primarily concerned with the clarification of concepts through the use of formal logic and mathematics. The systems scientist, for Churchman at least, is concerned with optimizing the operations of a more or less organized system of human beings, e.g., a business firm, a university, or a state. As Churchman sees it, there is little need for the formal philosopher within systems science because the attempt to become too precise too soon impedes progress toward the solution of the important problems at issue. On the other hand, there is, Churchman claims, a need for general criticism within systems science because of the danger that one may become so wrapped up in suboptimizing within a particular subsystem that he fails to see that in a larger context he is suboptimizing himself right out of existence. In the following brief comments I would like to suggest that there is no intrinsic connection between being a philosopher and being a critical systems scientist, but that there is a part of the philosophy of science that might correctly be described as a branch of systems science.

Many philosophers of science are concerned with the overall logical and mathematical structure of particular bodies of knowledge, e.g., relativity theory, quantum mechanics, learning theory, or perception. Studies of this sort are often referred to as foundational studies, and insofar as he is concerned solely with particular theories, there is little here to distinguish the philosopher from the theoretical scientist. But the philosopher is usually also interested in reaching more general conclusions about the structure of scientific knowledge and scientific concepts, conclusions that apply to several areas or perhaps even to science as a whole. Using a traditional expression, I will say that all such studies, specific or general, are concerned with the *logic of science*.

In addition to the logical structure of scientific knowledge, one might be interested in the process by which new knowledge may be acquired. In particular, one might be interested in discovering *efficient* methods of acquiring new scientific knowledge either in specific fields or in general. This latter study would seem to be a branch of systems science. Unfortunately it has not generally been recognized as a part of philosophy of science. This seems partly due to the mistaken inference that since there are no foolproof methods of acquiring knowledge, there are no objective methods at all. It seems also partly due to a misapplication of the distinction between discovery and justification. According to the conventional understanding of this distinction, the process of acquiring knowledge is something studied in the psychology and sociology of science – not in the philosophy of science. But clearly the search for *efficient* or *optimal* methods is not a task for psychology or sociology. Again drawing on a traditional phrase, I will say that the latter sorts of studies concern the *methodology of science*.

There is little recent literature in the philosophy of science that is explicitly methodological in the above sense. But much of the impact of philosophical investigations based on studies in the history of science seems due to their concern with general methodological principles. Unfortunately the failure to make a clear distinction between logic of science and methodology of science has lead many to exaggerate the amount of actual conflict between the recent historically based studies and traditional logical empiricist doctrines.

Churchman would no doubt argue that the distinction between the methodology of science and general systems science cannot be made. But clear examples come readily to mind. A methodological study of modern physics might lead to the conclusion that a giant particle accelerator should be built. This is clearly a case of suboptimization. A critical systems scientist at a higher level might conclude that the money would be better spent on low income housing in inner cities. But this latter conclusion clearly would not be part of the methodology of science and the critical systems scientist who drew it would not necessarily be a philosopher in any professional sense of this term.

At the moment I am inclined to think that Churchman should be satisfied if philosophers of science admit genuinely methodological studies as being part of the philosophy of science and that he should not go on to urge that critical systems science be allowed as well. A flexible discipline may be better than none at all. However, if Churchman feels that our overall situation has deteriorated so far that it is too late for the methodology of science, as conceived above, to do any good, then he should frankly argue that we forsake philosophy of science for general systems science. It is not inconceivable that one should be able to make a good case for this view. If it can be made, someone like Churchman ought to make it.

Indiana University

PETER C. FISHBURN

UTILITY THEORY WITH INEXACT PREFERENCES AND DEGREES OF PREFERENCE

ABSTRACT. $a - b \prec^* c - d$ is taken to mean that 'your' degree of preference for a over b is less than 'your' degree of preference for c over d. Various properties of the strength-of-preference comparison relation $\prec^*$ are examined along with properties of simple preferences defined from $\prec^*$. The investigation recognizes an individual's limited ability to make 'precise' judgments. Several utility theorems relating $a - b \prec^* c - d$ to $u(a) - u(b) < u(c) - u(d)$ are included.

I. INTRODUCTION

The concept of comparable preference differences ("my degree of preference for apple over orange exceeds my degree of preference for banana over pear") apparently received its first axiomatic expression at the hands of Pareto (1927, p. 264) and Frisch (1926). This concept continued to play a minor role in economic theory over the years, as evidenced for example in the papers by Lange (1934a, 1934b), Brown (1934), Bernardelli (1934), Allen (1934–5), Alt (1936), Samuelson (1938a), Armstrong (1939, 1948), and Weldon (1950). More recent axiomatizations of the notion of comparable preference differences, in the context of modern measurement theory, will be noted later.

The main purpose of this paper is to discuss the notion of 'inexact' comparisons of degrees of preference. With the exception of the paper by Adams (1965), the discussions of comparable preference difference have concentrated on an 'exact' notion, the realism of which is open to serious criticism. The material on degrees of preference will be preceded by a section on 'inexact' simple preference as characterized by the intransitivity of indifference, a notion that dates at least from Armstrong (1939).

Because the concept of comparable preference differences (degree of preference, strength of preference, preference intensity) has been intertwined with the phrase 'measurable utility' I shall comment on this phrase and two of its cousins in the final section of the paper.

II. SIMPLE PREFERENCE COMPARISONS

Before examining comparable preference differences I shall take a moment to review some theory of simple preference comparisons. We shall consider a simple preference relation $\prec$ on a set A, with $a \prec b$ interpreted as 'a is worse than b' or 'b is preferred to a'. Most studies of simple preference assume that $\prec$ is a *strict partial order*, being irreflexive ($a \prec a$ is false) and transitive ($a \prec b$ and $b \prec c$ imply $a \prec c$). Then $\prec$ is asymmetric also ($a \prec b$ implies not $b \prec a$). Defining *indifference* $\sim$ as the absence of strict preference, so that

(1) $a \sim b$ if and only if not $a \prec b$ & not $b \prec a$,

$\sim$ is reflexive ($a \sim a$) and symmetric ($a \sim b$ implies $b \sim a$) when $\prec$ is irreflexive. Indifference need not be transitive when $\prec$ is a strict partial order, for $a \sim b$ & $b \sim c$ & $a \prec c$ do not violate the axioms for a strict partial order.

When $\sim$ is assumed to be transitive and $\prec$ is a strict partial order, we shall refer to $\prec$ as a *weak order*. It is easily verified that $\prec$ is a weak order if and only if it is asymmetric and is negatively transitive: not $a \prec b$ & not $b \prec c \Rightarrow$ not $a \prec c$. The term 'negative transitivity' is used by Chipman (1971) and is equivalent to: if $a \prec b$ then, for any $c \in A$, either $a \prec c$ or $c \prec b$.

Until fairly recently, studies in preference theory usually assumed that $\prec$ is a weak order. The earliest study I know of in which the transitivity of indifference was seriously challenged was done by Armstrong (1939). This and later work by Armstrong (1948, 1950) did not appear to have a great impact until Luce (1956) incorporated the notion of intransitive indifference in his study of semiorders. In the last few years, due largely to the pioneering of Armstrong and Luce, a number of papers on preference theory without transitive indifference have appeared. Since Fishburn (1970a) provides a survey of this literature I shall not review it in detail here.

'Inexact preferences' in the title of this paper refers mainly to the possibility of intransitive indifference, which, to quote Armstrong (1950, p. 122), arises from "the imperfect powers of discrimination of the human mind whereby inequalities become recognizable only when of sufficient magnitude." For example, an individual may be indifferent to small and perhaps undetectable changes (consider adding one grain of sugar to a cup of coffee), but when enough of these changes are placed end-to-end the difference between the initial and the final positions will be most noticeable. For another example suppose a, b, and c are respectively the alternatives 'get \$138', 'get \$139', and 'flip a coin then get \$0 if it lands heads up and \$390 if it lands tails up.' Surely $a \prec b$ for most people, but $a \sim c$ and $b \sim c$ also do not seem unreasonable.

Among the possible real-valued utility representations for simple preferences, the following two are the most common:

(2) $\quad a \prec b \Rightarrow u(a) < u(b)$,
(3) $\quad a \prec b \Leftrightarrow u(a) < u(b)$,

where, as usual, $\Rightarrow$ means 'implies' and $\Leftrightarrow$ means 'if and only if'.

When A is countable (finite or denumerable), (3) holds for all $a, b \in A$ if and only if $\prec$ on A is a weak order, and (2) holds for all $a, b \in A$ when $\prec$ is a strict partial order. [The strict partial order assumptions are not fully necessary for (2). All that is required is that the transitive closure of $\prec$ be asymmetric, which permits intransitive preference of the form $a \prec b$ & $b \prec c$ & $a \sim c$ but prohibits an intransitive preference cycle such as $a_1 \prec a_2$ & $a_2 \prec a_3$ & ... & $a_{n-1} \prec a_n$ & $a_n \prec a_1$.] Additional assumptions, noted by Fishburn (1970a, 1970b), may be required for (2) and (3) when A is uncountable.

Suppose $a \sim b$ & $b \sim c$ & $a \prec c$, so that $\sim$ is not transitive. Then, if (2) applies, $u(a) < u(c)$ and it is not possible to have both $u(b) = u(a)$ and $u(b) = u(c)$. Thus, when indifference is not transitive, orderpreserving utilities as in (2) cannot be 'indifference-preserving'. When (3) holds, $a \sim b \Leftrightarrow u(a) = u(b)$.

Semiorders

The most popular intermediate between a strict partial order and a weak order in preference theory is the semiorder introduced by Luce (1956). Using the definition of Scott and Suppes (1958), $\prec$ on A is a semiorder when it is irreflexive and satisfies

(4) $\quad a \prec b$ & $c \prec d \Rightarrow a \prec d$ or $c \prec b$,
(5) $\quad a \prec b$ & $b \prec c \Rightarrow a \prec d$ or $d \prec c$.

Either (4) or (5) along with irreflexivity implies that $\prec$ is transitive.

Hence a semiorder is a strict partial order although the converse is false. Although a semiorder permits intransitive indifference and hence need not be a weak order it restricts such intransitivity in ways not done by a strict partial order. Suppose for example that $a \sim b$ & $b \sim c$ & $a \prec c$. Then, when $\prec$ is a strict partial order, we can have $a \prec d \prec c$ although this is prohibited by (5).

In cases where preference does not decrease (or else does not increase) along only one underlying factor, the assumptions of a semiorder seem about as reasonable as those of a strict partial order. However, most situations where preference is of interest are not of this sort and then the semiorder axioms (4) and (5) are open to serious criticism. Suppose, for example, that with a, b, and c as 'win \$138', 'win \$139', and 'flip a coin and win \$0... or \$390...', we take $d =$ 'win \$138.50' and $e =$ 'flip a coin and win \$0... or \$391...'. With $a \sim c$ & $b \sim c$ & $a \prec b$, it would not seem unreasonable to have $a \prec d \prec b$ and $d \sim c$, which violates (5), or to have $a \sim e$ & $b \sim e$ & $c \prec e$, which violates (4). Since risk is a factor along with money in this example, the example does not fall into the category of only one underlying factor.

Scott and Suppes (1958) prove that when A is finite and $\prec$ is a semiorder then there is a real-valued function u on A such that, for all $a, b \in A$,

(6) $a \prec b \Leftrightarrow u(a) + 1 < u(b)$.

The constant in (6) represents a just noticeable difference along the utility scale used for $\prec$. The utility representation (6) is an intermediary between (2) and (3).

Following Luce (1956, p. 182), we shall consider a relation $\prec^+$ defined by

(7) $a \prec^+ b \Leftrightarrow (a \sim c$ & $c \prec b)$ or $(a \prec c$ & $c \sim b)$ for some $c \in A$.

$a \prec b \Rightarrow a \prec^+ b$ as long as $\prec$ is irreflexive. You may find it instructive to verify that $\prec^+$ is a weak order when $\prec$ is a semiorder, and hence if A is countable then (3) applies with $\prec^+$ in place of $\prec$ when $\prec$ is a semiorder. Conversely, it is easily shown that $\prec$ is a semiorder when $\prec$ is a strict partial order and $\prec^+$ is asymmetric. An immediate corollary of these facts is that $\prec^+$ is a weak order when it is asymmetric and $\prec$ is a strict partial order.

A Return to Strict Partial Orders

Because $\prec^+$ is not very useful by itself unless it is asymmetric, we consider a form of 'induced preference' that does not depend on $\prec$ being a semiorder for its asymmetry. Specifically, let

(8) $\quad a \prec^\circ b \Leftrightarrow a \prec^+ b \ \& \ \text{not } b \prec^+ a,$

which is asymmetric by definition. Moreover, when $\prec$ is a strict partial order, $a \prec b \Rightarrow a \prec^\circ b \Rightarrow a \prec^+ b$, although the reverse implications can fail. If $A = \{a, b, c\}$ and $a \sim b \ \& \ b \sim c \ \& \ a \prec c$, then $a \prec^\circ b \prec^\circ c$.

$a \prec^\circ b$ indicates that $a \prec b$ or else that $a \sim b$ and [as seen through intermediary elements in A as in (7) and (8)] a can be viewed as 'slightly inferior' to b although the two are indifferent in the sense of (1). When $\prec$ is a semiorder, $a \prec^\circ b \Leftrightarrow a \prec^+ b$.

In the remainder of this section we shall prove that the induced preference relation $\prec^\circ$ is a strict partial order when $\prec$ is a strict partial order. Since $\prec^\circ$ is asymmetric it is irreflexive. To prove that $\prec^\circ$ is transitive when $\prec$ is a strict partial order suppose to the contrary that $a \prec^\circ b \ \& \ b \prec^\circ c \ \& \ \text{not } a \prec^\circ c$. Now not $a \prec^\circ c$ means that either

(i) $\quad c \sim x \ \& \ x \prec a$ for some $x \in A$, or

(ii) $\quad c \prec y \ \& \ y \sim a$ for some $y \in A$, or

(iii) $\quad (a \sim p \ \& \ p \prec c)$ or $(a \prec p \ \& \ p \sim c)$ for no $p \in A$.

Suppose that (i) holds. Then not $b \prec^+ a$ requires $x \prec b$, but this along with $c \sim x$ implies that $c \prec^+ b$, which is false by hypothesis. Likewise, if (ii) holds then not $c \prec^+ b$ requires $b \prec y$, which along with $y \sim a$ implies $b \prec^+ a$, which contradicts $a \prec^\circ b$. Finally, suppose that (iii) holds, and for $a \prec^\circ b$ assume first that $a \sim r \ \& \ r \prec b$. Then, by (iii), not $r \prec c$, and hence either $c \prec r$ or $c \sim r$. But $c \sim r$ along with $r \prec b$ implies $c \prec^+ b$, a contradiction to $b \prec^\circ c$, and $c \prec r$ implies $c \prec b$ by transitivity, again contradicting $b \prec^\circ c$. If for $a \prec^\circ b$ we assume that $a \prec s \ \& \ s \sim b$ then, by (iii), not $s \sim c$. But if $s \prec c$ then $a \prec c$, which contradicts not $a \prec^\circ c$, and if $c \prec s$ then $c \prec^+ b$, which contradicts $b \prec^\circ c$. Therefore not $a \prec^\circ c$ must be false when $a \prec^\circ b \ \& \ b \prec^\circ c$ and $\prec$ is a strict partial order.

III. COMPARABLE PREFERENCE DIFFERENCES

The view of preference in this paper is essentially subjective. Introspection is taken to be the basic means by which an individual recognizes a preference. A generalization of this notion to a second-order form of comparison suggests that an individual may be able to perceive by introspection that his degree of preference for a over b exceeds his degree of preference for c over d. For example, a person may prefer three candidates a, b, and c in a major political contest in that order and feel that his degree of preference for a over b exceeds that for b over c. He feels strongly about wanting to see a elected, and if a is not elected he doesn't much care whether b or c is, although he has a slight preference for b over c. For another example, a poor man might find that he would get more incremental joy from a (tax-free) gift of $500 as compared to a gift of $1 than he would from a gift of $1 020 000 as compared to a gift of $1 000 000.

Although these examples illustrate the notion of comparing preference differences (or degrees of preference, strength of preference, intensity of preference), the viewpoint taken here is that, like simple preference, degree of preference is essentially unanalyzable in terms of other constructs. Because of its supposedly nonoperational character, a number of people view degree of preference as a sterile concept. This is especially true in economic consumption theory where the introspective notion of preference has been largely replaced by analyses based on actual or supposed choices. This development was stimulated by Samuelson (1938b) who "tried here to develop the theory of consumer's behaviour freed from any vestigial traces of the utility concept" (p. 71). Later connections between Samuelson's concept of 'revealed preference' (revealed through choices) and preference as considered here are discussed by Arrow (1959), Houthakker (1961), Richter (1966), Hansson (1968), and Hurwicz and Richter (1971), among others.

Before going on to a technical examination of degree of preference I should note that several authors feel that strength of preference can be measured using constructs wholly based on the notion of simple preference. This point of view, which has been repudiated by other authors, will be examined at the end of this paper.

Degree of Preference

With A a set of alternatives we shall let $\prec^*$ be a binary relation on pairs of ordered pairs (a, b) in the product set $A \times A$. $(a, b) \prec^* (c, d)$ is interpreted as "the degree of preference for a over b is less than the degree of preference for c over d." Hoping for some gain in clarity I shall use $a-b$ as an alternative way of writing the ordered pair (a, b). Thus $a-b \prec^* c-d \Leftrightarrow (a, b) \prec^* (c, d)$. Two standard assumptions for $\prec^*$ that bring out the directed difference notion are

(9) $\quad a-b \prec^* c-d \Rightarrow d-c \prec^* b-a,$
(10) $\quad a-b \prec^* c-d \Rightarrow a-c \prec^* b-d.$

Suppose a is preferred to b, and c is preferred to d. Then both $a-b$ and $c-d$ would be expected to be 'positive' and $a-b \prec^* c-d$ means that $c-d$ is the larger preference difference. Hence $d-c$ would be a greater 'negative' difference than $b-a$, and this is reflected by (9).

Assumption (10) is illustrated on a line of increasing preference that preserves the comparable preference differences:

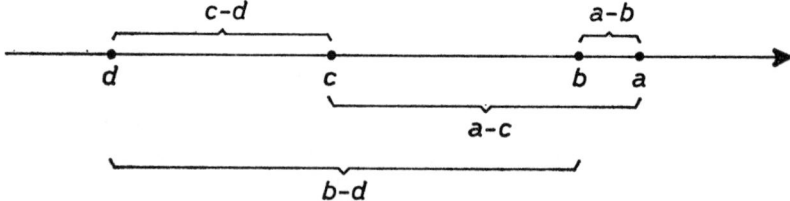

The difference $b-c$ in this picture is common to both $a-c$ and $b-d$, and when $b-c$ is 'subtracted' from these two we are left with the original $a-b$ and $c-d$.

Since I see no reason why an individual should be any more precise in his judgments of preference differences than he is in his simple preference judgments, it seems natural to regard $\prec^*$ as a strict partial order (irreflexive, transitive) so that $\sim^*$, defined by

(11) $\quad a-b \sim^* c-d \Leftrightarrow$ not $a-b \prec^* c-d$ & not $c-d \prec^* a-b,$

is not assumed to be transitive. Naturally, if $\sim^*$ were assumed to be transitive along with $\prec^*$ a strict partial order then $\prec^*$ would be a weak order. To illustrate the intransitivity of $\sim^*$, our poor man might have \$500 − \$10 $\sim^*$ \$500 000 − \$400 000 and \$500 − \$0 $\sim^*$ \$500 000 − \$400 000 along with \$500 − \$10 $\prec^*$ \$500 − \$0. For most people I suspect that x in \$100 − \$$x$ $\sim^*$ \$$x$ − \$0 would cover a range of possible values rather than a single point.

As shown by the second part of the following lemma, (9) and (10) for $\prec^*$ imply similar properties for $\sim^*$.

LEMMA 1. $[\prec^*$ *is irreflexive* & $(10)] \Rightarrow a - a \sim^* b - b$. $[(9) \& (10)] \Rightarrow$
$\Rightarrow (a - b \sim^* c - d \Leftrightarrow a - c \sim^* b - d \Leftrightarrow d - c \sim^* b - a)$.

Contrary to the conclusion of the first part suppose that $a - a \prec^* b - b$. Then, by (10), $a - b \prec^* a - b$, which contradicts the irreflexivity of $\prec^*$. For the latter part of the lemma, $a - b \sim^* c - d \Rightarrow$ not $a - b \prec^* c - d \Rightarrow$ not $a - c \prec^* b - d$ [by (10)] $\Rightarrow (a - c \sim^* b - d$ or $b - d \prec^* a - c)$ the latter of which gives $c - a \prec^* d - b$ by (9). But also $a - b \sim^* c - d \Rightarrow$ not $c - d \prec^* a - b \Rightarrow$ not $c - a \prec^* d - b$ by (10). Therefore $a - b \sim^* c - d \Rightarrow$
$\Rightarrow a - c \sim^* b - d$. Proof of the final implication in the lemma is left to the reader.

A further connection between (9), (10), and $\sim^*$ follows.

LEMMA 2. $[\prec^*$ *is asymmetric* & $(10) \& (a - b \sim^* c - d \Leftrightarrow a - c \sim^*$
$\sim^* b - d)] \Rightarrow (9)$.

Contrary to (9) suppose that $a - b \prec^* c - d$ & not $d - c \prec^* b - a$. By the first of these and (10), $a - c \prec^* b - d$. By the latter, either $b - a \prec^*$
$\prec^* d - c$ or $b - a \sim^* d - c$, so that with (10) or the $\sim^*$ hypothesis of the lemma, $b - d \prec^* a - c$ or $b - d \sim^* a - c$. Then $a - c \prec^* b - d$ and $b - d \prec^* a - c$ contradict asymmetry, and $a - c \prec^* b - d$ and $a - c \sim^*$
$\sim^* b - d$ contradict (11). Thus Lemma 2 is proved.

Connection with Simple Preference

Of the several ways that one might relate $\prec$ to $\prec^*$, the following definition of $\prec$ in terms of $\prec^*$ is fairly standard:

(12) $a \prec b \Leftrightarrow a - a \prec^* b - a$.

With $a - a$ interpreted as 'zero' difference, (12) says that b is preferred to a if and only if the preference difference $b - a$ is 'positive'. At least one person, Armstrong (1939), does not agree with (12). His somewhat unusual viewpoint is examined at the end of this section. In the meantime we shall use (12).

Because of the obvious interpretation of $a - a$ it is natural to assume that

(13) $a - a \prec^* b - a \Rightarrow c - c \prec^* b - a$ for all $c \in A$.

When $a \prec b$ by (12) it is natural to expect that $c - b \prec^* c - a$ for any

$c \in A$. In fact, if (10) and (13) hold, we have $a \prec b \Rightarrow a - a \prec^* b - a \Rightarrow$
$\Rightarrow c - c \prec^* b - a \Rightarrow c - b \prec^* c - a$. Also, by (9), $a - c \prec^* b - c$. A closely related result is

LEMMA 3. *[$\prec^*$ is a strict partial order & (10) & (13)] $\Rightarrow \prec$ as defined by (12) is a strict partial order.*

Under the hypotheses of the lemma suppose that $a \prec b$ and $b \prec c$. Then, by (12) and (13), $a - a \prec^* b - a$ and $a - a \prec^* c - b$. Using (10) on both of these we get $a - b \prec^* a - a$ and $a - c \prec^* a - b$ so that $a - c \prec^* a - a$ by the transitivy of $\prec^*$. Hence, by (10), $a - a \prec^* c - a$, or $a \prec c$. This proves that $\prec$ is transitive. It is irreflexive also since $\prec^*$ is irreflexive.

The hypotheses of Lemma 3 do not of course specify the minimal conditions on $\prec^*$ that are necessary and sufficient for $\prec$ to be a strict partial order. (These conditions are obviously not $a - a \prec^* a - a$, and $[a - a \prec^* b - a$ & $b - b \prec^* c - b] \Rightarrow a - a \prec^* c - a$.) Rather the purpose of Lemma 3 is to show that rather natural assumptions on $\prec^*$ lead to $\prec$ being a strict partial order.

In a similar vein it is easily seen that it is not necessary to assume that $\prec^*$ is a weak order in order to prove that $\prec$ is a weak order. But whatever assumptions might be used to obtain $\prec$ as a weak order would be liable to the same kind of criticism that leads to the viewpoint that indifference $\sim$ ought not to be regarded as transitive.

Utility Representations

In terms of a utility function it is natural to associate $a - b \prec^* c - d$ with $u(a) - u(b) < u(c) - u(d)$. An alternative approach is used by Suppes and Winet (1955) who work with absolute (undirected) difference comparisons and associate the degree of preference between a and b with $|u(a) - u(b)|$.

Corresponding to (2) and (3) for simple preference, two obvious difference representations are

(14) $a - b \prec^* c - d \Rightarrow u(a) - u(b) < u(c) - u(d)$
(15) $a - b \prec^* c - d \Leftrightarrow u(a) - u(b) < u(c) - u(d)$.

Under (12), (2) follows from (14), and (3) follows from (15). Although (15) requires $\prec^*$ to be a weak order, (14) does not imply that $\prec^*$ is a strict partial order (although it does require that the transitive closure of $\prec^*$ be asymmetric and hence a strict partial order).

In no event are the assumptions for $\prec^*$ noted above sufficient for either (14) or (15). For example, if

(16) $\quad a - b \prec^* c - d$
(17) $\quad e - a \prec^* d - f$

then $u(e) - u(b) < u(c) - u(f)$ by either (14) or (15), so that for (14) to hold we require not $c - f \prec^* e - b$, and for (15) to hold we require $e - b \prec^* c - f$. Neither of these is implied by $\prec^*$ being a weak order along with (9), (10), (16), and (17). See also Samuelson (1938a) on this point.

By extending the idea of the preceding paragraph we can obtain conditions on $\prec^*$ that are necessary and sufficient for (14), or for (15), when A is a finite set. The necessary and sufficient conditions for (14) are given by Adams (1965). They are:

(18) $\quad [a_1, ..., a_m, d_1, ..., d_m$ is a permutation (reordering) of $b_1, ..., b_m, c_1, ..., c_m$ & $a_k - b_k \prec^* c_k - d_k$ for all $k < m] \Rightarrow$ not $a_m - b_m \prec^* c_m - d_m$.

This is to hold for $m = 2, 3, \ldots$. The necessary and sufficient conditions for (15) [with A finite] are given by Scott (1964). Scott's conditions are equivalent to:

(19) $\quad [a_1, ..., a_m, d_1, ..., d_m$ is a permutation (reordering) of $b_1, ..., b_m, c_1, ..., c_m$ & $a_k - b_k \prec^* c_k - d_k$ or $a_k - b_k \sim^* c_k - d_k$ for each $k < m] \Rightarrow$ not $a_m - b_m \prec^* c_m - d_m$.

Proofs of the theorems for (14) and (15) are also contained in Fishburn (1970b, Chapter 6).

You can easily prove that (18) does not imply either (9) or (10), which are not required for (14). On the other hand, (19) implies (9), (10), and that $\prec^*$ is a weak order.

Another difference representation for $\prec^*$ that resembles (6) for a semiorder $\prec$ and is intermediate between (14) and (15) is

(20) $\quad a - b \prec^* c - d \Leftrightarrow u(a) - u(b) + 1 < u(c) - u(d)$.

Using the Theorem of The Alternative much as it is used in establishing (14) and (15) in the references cited above, it is not hard to show that

(20) holds when A is finite if and only if $\prec^*$ is irreflexive and satisfies the following relative of (18) and (19):

(21) $\quad [a_1, ..., a_{2m}, d_1, ..., d_{2m}$ is a permutation of $b_1, ..., b_{2m}$, $c_1, ..., c_{2m}$ & $a_k - b_k \sim^* c_k - d_k$ for $k = 1, ..., m$ & $a_k - b_k \prec^*$ $\prec^* c_k - d_k$ for $k = m+1, ..., 2m-1] \Rightarrow$ not $a_{2m} - b_{2m} \prec^* c_{2m} - d_{2m}$.

This is to hold for all positive integers m. Since it and irreflexivity imply (20), the semiorder conditions for $\prec^*$ must follow from (21).

Sufficient conditions on $\prec^*$ for the exact representation (15) when A is infinite are given by Suppes and Winet (1955), Scott and Suppes (1958) [see also Suppes and Zinnes (1963, pp. 34–38)], Pfanzagl (1959, 1968), Chipman (1960), and Luce (1968). As noted before, Suppes and Winet take the route of absolute difference comparisons. Luce (1968) works with 'positive' differences. A slight twist in one of Debreu's topologically-oriented axiom systems (1960) also gives a set of conditions sufficient for (15). This is discussed by Fishburn (1970b, Chapter 6) along with several of the other theories cited in this paragraph.

To the best of my knowledge there does not presently exist a reasonable axiomatization for the inexact representation (14) when A is allowed to be infinite. (Naturally, the theories of the preceding paragraph imply (14), but I am here thinking of axioms for (14) that do not suppose that $\prec^*$ is a weak order.)

Although there are differences in the infinite A axioms used to arrive at (15), each of the theories cited above for infinite A implies that u is unique up to a positive linear transformation. This means that a real-valued function v on A satisfies (15) in place of u if and only if there are real numbers r and s with $r > 0$ and

(22) $\quad v(a) = r u(a) + s \quad$ for all $\quad a \in A$.

It should be noted that (15) alone does not imply that u is unique up to a positive linear transformation, or unique except for origin and scale unit. The uniqueness depends also on fairly strong structural assumptions for A and $\prec^*$ that are not essential for (15).

This can be illustrated by the usual bisection scaling method that would apply for the exact theories. To simplify things a bit, suppose that A contains a worst alternative w and a best alternative b. It is then supposed that there is a third alternative m that is precisely midway in preference between w and b with $m - w \sim^* b - m$, so that $u(m) = [u(w) + u(b)]/2$. The process continues by determining a midpoint between w and m, and another between m and b. With $u(w) = 0$ and $u(b) = 1$, the continuation of this process yields a set of utility values that is dense in the interval from 0 to 1. Under appropriate assumptions on A and $\prec^*$, the utilities for alternatives not assigned in this construction process (and not indifferent to an alternative appearing in the construction) can be squeezed into the interval [0, 1] so that (15) holds throughout A.

Armstrong's Unusual Viewpoint

Armstrong (1939) advocates precise measurement for u, using a bisection argument like that in the preceding paragraph. Put another way, he advocates (15) with u therein unique up to a positive linear transformation as in (22). At the same time he argues that, when $u(a) < u(b)$, we should not have $a \prec b$ unless the difference $u(b) - u(a)$ is as great as some minimal positive threshold level. In other words, in supporting intransitive indifference ($\sim$), Armstrong proposes a semiorder representation like (6) for $\prec$.

This viewpoint is clearly contrary to the definition of $\prec$ from $\prec^*$ given by (12). Although this in itself might not seem alarming, closer examination suggests that Armstrong's argument is not wholly coherent. The difficulty, as I see it, is as follows. Suppose in fact that, with w and b as given above, one is indeed able to identify an m that is precisely midway in preference between w and b. As this process continues, the preference intervals whose midpoints are to be determined become ever narrower, and (to hold to Armstrong's semiorder view for $\prec$) must eventually reach a position where the endpoints of a subinterval whose midpoint is to be determined are in fact indifferent to each other. If these endpoints are indifferent, how is it possible to locate a point that is precisely midway in preference between the end points?

Perhaps it is not surprising that the viewpoint put forth by Armstrong has not been widely adopted.

IV. ORDINAL, CARDINAL AND MEASURABLE UTILITY

In concluding this discussion, a few words about the terms that title this section are in order. Because these terms have been applied almost exclusively to 'exact' utility models like (3) and (15), I shall concentrate on these.

Ordinal Utility

'Ordinal utility' is used to denote a function that preserves the preference order $\prec$ as in (3), in which case the set of indifference ($\sim$) classes in A is order isomorphic to a subset of the real numbers. When (3) holds, all order-preserving transformations of u satisfy (3) also. Hence if $u(a) < u(b) < u(c)$, no special significance is attached to the differences $u(b) - u(a)$ and $u(c) - u(b)$ since an order-preserving transformation of u can reverse the order of such differences.

The term ordinal seems also to have been used for a qualitative concept of utility. Thus, one might say that *preferences* are ordinal when $\prec$ on A is a weak order. This gives rise to the so-called indifference map approach in economic theory. In this context, where A is the non-negative orthant of n-dimensional Euclidean space, additional assumptions (as noted previously) are needed to insure the existence of an ordinal utility function as in (3).

In addition to the weak order axiom, other axioms for $\prec$ on A may give rise to properties for u (e.g., continuity, convexity) beyond those found in (3). A broad notion of ordinal utility would include such systems, provided that they are based only on simple preference and not on the degree of preference relation $\prec^*$.

Cardinal and Measurable Utility

In recent years 'cardinal utility' and 'measurable utility' have been used to denote a real-valued function u on A that satisfies not only (3) but also satisfies some additional condition, say C, such that, when u satisfies (3) and C along with appropriate structural properties, it is unique up to a positive linear transformation as in (22). As Strotz (1953), Chipman (1960), and others have pointed out, any order-preserving transformation of u will continue to satisfy (3) in such cases but will satisfy C also only if it is linear. We shall consider three examples.

1. C is (15): $a - b \prec^* c - d \Leftrightarrow u(a) - u(b) < u(c) - u(d)$. As noted above, appropriately strong assumptions on $\prec^*$ will yield (15), and hence (3) by way of (12), with u satisfying (3) and (15) unique up to a positive linear transformation.

2. Let A be the set of all probability distributions on a finite set X. In this case the von Neumann and Morgenstern (1947) axioms for $\prec$ on A give rise to a real-valued function u on A that, for all $a, b \in A$ and $0 \leq \alpha \leq 1$, satisfies (3) and

(23) $\quad u(\alpha a + (1-\alpha) b) = \alpha u(a) + (1-\alpha) u(b),$

in which $\alpha a + (1-\alpha) b \in A$ is the direct linear combination of a and b. C is (23) in this case. When u is extended to X by defining $u(x) = u(a)$ when $a(x) = 1$, (23) gives the expected-utility form $u(a) = \sum_x a(x) u(x)$. When u satisfies (3) and (23), it is unique up to a positive linear transformation.

3. Let A be the set $A_1 \times A_2$ of all ordered pairs $a = (a_1, a_2)$ with $a_1 \in A_1$ and $a_2 \in A_2$. Under appropriately strong conditions for $\prec$ on A, Debreu (1960), Luce and Tukey (1964), Krantz (1964, 1967) and Luce (1966), derive the existence of real-valued functions u, u_1, and u_2 on A, A_1, and A_2 respectively that satisfy (3) along with

(24) $\quad u(a_1, a_2) = u_1(a_1) + u_2(a_2),$

such that v, v_1 and v_2 on A, A_1, and A_2 satisfy (3) and (24) also if and only if there are numbers r, s_1, and s_2 with $r > 0$ such that $v(a) = ru(a) + (s_1 + s_2)$, $v_1(a_1) = ru_1(a_1) + s_1$, and $v_2(a_2) = ru_2(a_2) + s_2$. In this case C is (24).

We repeat again that any order-preserving transformation of u in any of these examples will satisfy (3) but will not satisfy C unless it is linear. For example, in the expected-utility example we are not forced into using (23) when (3) holds, although (23) is useful since computation by expectation is mathematically convenient.

Examples 2 and 3 differ from 1 in that the axioms for (23) and (24) are based entirely on simple preference comparisons ($\prec$) and not on the notion of a higher-order degree of preference relation. Using the broad notion of ordinal utility explained above, Baumol (1958) thus speaks of u in example 2 as 'the cardinal utility which is ordinal'. Example 3 for additive utilities is another case of a 'cardinal utility which is ordinal'.

From Stigler's history of utility (1950) it appears that the notions of measurable utility that arose among the founders of utility theory in the second half of the nineteenth century were not unlike the characterization given above, with the exception that a few writers regarded utility as absolutely measurable (with a natural zero and an arbitrary unit). As both Stigler (1950) and Strotz (1953) note, the early views were largely introspective: utility was "thought of as a psychological entity measurable in its own right" (Strotz, p. 384).

This does not mean however that founders such as Jevons and Walras used the Pareto-Frisch notion of comparable preference differences explicitly in their theories. In fact they used additive functional forms for utility on commodity bundles like that in example 3, and, as we have noted, this form can be derived from purely ordinal ($\prec$) assumptions. This was clearly recognized by Fisher (1892, 1925) who constructed the utility functions for different commodities, under the independence assumptions required for (24), by considering trade-offs between pairs of commodities.

Later on the introspective notion of 'measurable utility' lost ground in economics when writers such as Slutsky (1915) and Hicks and Allen (1934) showed that 'measurable utility' was not required to obtain certain desired results in the static theory of consumer demand. As noted before, even the introspective approach to simple preferences has been (at least partly) set aside in consumption theory by the work of Samuelson (1938b) and others. [Further history on this point is given by Houthakker (1961).] Nevertheless Frisch (1926, 1959, 1964) has steadfastly argued for the usefulness of some notion of measurable utility in connection, for example, with time-dependent consumer demand.

A Final Note

I should like to conclude with a point made by many other writers. Some people apparently have the impression that a utility function u that (because of certain properties it is assumed to satisfy) is unique up to a positive linear transformation is necessarily an accurate measure of preference intensities or degrees of preference. As Weldon (1950), Strotz (1953), Ellsberg (1954), Luce and Raiffa (1957, p. 32), Baumol (1958), Arrow (1963, pp. 9-11) and I don't know how many others have cautioned, there is nothing in the theories of cases like examples 2 and 3 that justifies this impression since they are based wholly on simple preference comparisons.

Research Analysis Corporation

BIBLIOGRAPHY

Adams, E. W., 'Elements of a Theory of Inexact Measurement', *Philosophy of Science* **32** (1965) 205–28.
Allen, R. G. D., 'A Note on the Determinateness of the Utility Function', *Review of Economic Studies* **2** (1934–35) 155–8.
Alt, F., 'Über die Messbarkeit des Nutzens', *Zeitschrift für Nationalökonomie* **7** (1936) 161–9.
Armstrong, W. E., 'The Determinateness of the Utility Function', *Economic Journal* **49** (1939) 453–67.
Armstrong, W. E., 'Uncertainty and the Utility Function', *Economic Journal* **58** (1948) 1–10.
Armstrong, W. E., 'A Note on the Theory of Consumer's Behaviour', *Oxford Economic Papers*, N.S. **2** (1950) 119–22.
Arrow, K. J., 'Rational Choice Functions and Orderings', *Economica*, N.S. **26** (1959) 121–7.
Arrow, K. J., *Social Choice and Individual Values*, 2nd ed., John Wiley and Sons, New York, 1963.
Baumol, W. J., 'The Cardinal Utility Which is Ordinal', *Economic Journal* **68** (1958) 665–72.
Bernardelli, H., 'Notes on the Determinateness of the Utility Function', *Review of Economic Studies* **2** (1934) 69–75.
Brown, E. H. P., 'Notes on the Determinateness of the Utility Function', *Review of Economic Studies* **2** (1934) 66–9.
Chipman, J. S., 'The Foundations of Utility', *Econometrica* **28** (1960) 193–224.
Chipman, J. S., 'Consumption Theory Without Transitive Indifference', in *Preferences, Utility, and Demand* (ed. by J. S. Chipman, L. Hurwicz, M. K. Richter, and H. Sonnenschein), Harcourt Brace/Jovanovich, 1971.
Debreu, G., 'Topological Methods in Cardinal Utility Theory', in *Mathematical Methods in The Social Sciences, 1959* (ed. by K. J. Arrow, S. Karlin, and P. Suppes), Stanford University Press, Stanford, California, 1960.
Ellsberg, D., 'Classic and Current Notions of "Measurable Utility"', *Economic Journal* **64** (1954) 528–56.
Fishburn, P. C., 'Intransitive Indifference in Preference Theory: A Survey', *Operations Research* **18** (1970a) 207–28.
Fishburn, P. C., *Utility Theory for Decision Making*, John Wiley and Sons, New York, 1970b.
Fisher, I., *Mathematical Investigations in the Theory of Value and Prices*, Yale University Press, New Haven, Connecticut, 1925. (Reprint of 1892 edition.)
Frisch, R., 'Sur une problème d'économie pure', *Norsk Mathematisk Forenings Skrifter*, Seril 1, **16** (1926) 1–40.
Frisch, R., 'A Complete Scheme for Computing All Direct and Cross Demand Elasticities in a Model with Many Sectors', *Econometrica* **27** (1959) 177–96.
Frisch, R., 'Dynamic Utility', *Econometrica* **32** (1964) 418–24.
Hansson, B., 'Choice Structures and Preference Relations', *Synthese* **18** (1968) 443–58.
Hicks, J. R. and Allen, R. G. D., 'A Reconsideration of the Theory of Value', *Economica*, N.S. **1** (1934) 52–76 and 196–219.
Houthakker, H.S., 'The Present State of Consumption Theory', *Econometrica* **29** (1961) 704–40.

Hurwicz, L. and Richter, M. K., 'Revealed Preference Without Demand Continuity Assumptions', in *Preferences, Utility, and Demand* (ed. by J. S. Chipman, L. Hurwicz, M. K. Richter, and H. Sonnenschein), Harcourt/Brace/Jovanovich, 1971.

Krantz, D. H., 'Conjoint Measurement: The Luce-Tukey Axiomatization and Some Extensions', *Journal of Mathematical Psychology* **1** (1964) 248–77.

Krantz, D. H., 'A Survey of Measurement Theory', *Michigan Mathematical Psychology Program* MMPP 67-4, University of Michigan, Ann Arbor, 1967.

Lange, O., 'The Determinateness of the Utility Function', *Review of Economic Studies* **1** (1934a) 218–25.

Lange, O., 'Notes on the Determinateness of the Utility Function', *Review of Economic Studies* **2** (1934b) 75–7.

Luce, R. D., 'Semiorders and a Theory of Utility Discrimination', *Econometrica* **24** (1956) 178–91.

Luce, R. D., 'Two Extensions of Conjoint Measurement', *Journal of Mathematical Psychology* **3** (1966) 348–70.

Luce, R. D., 'On the Numerical Representation of Qualitative Conditional Probability', *Annals of Mathematical Statistics* **39** (1968) 481–91.

Luce, R. D. and Raiffa, H., *Games and Decisions*, John Wiley and Sons, New York, 1957.

Luce, R. D. and Tukey, J. W., 'Simultaneous Conjoint Measurement: A New Type of Fundamental Measurement', *Journal of Mathematical Psychology* **1** (1964) 1–27.

Pareto, V., *Manuel d'Economie Politique*, 2nd ed., Marcel Giard, Paris, 1927.

Pfanzagl, J., 'A General Theory of Measurement: Applications to Utility', *Naval Research Logistics Quarterly* **6** (1959) 283–94.

Pfanzagl, J. (in cooperation with V. Baumann and H. Huber), *Theory of Measurement*, John Wiley and Sons, New York, 1968.

Richter, M. K., 'Revealed Preference Theory', *Econometrica* **34** (1966) 635–45.

Samuelson, P. A., 'The Numerical Representation of Ordered Classifications and the Concept of Utility', *Review of Economic Studies* **6** (1938a) 65–70.

Samuelson, P. A, 'A Note on the Pure Theory of Consumer's Behaviour', *Economica*, N.S. **5** (1938b) 61–71 and 353–4.

Scott, D., 'Measurement Structures and Linear Inequalities', *Journal of Mathematical Psychology* **1** (1964) 233–47.

Scott, D. and Suppes, P., 'Foundational Aspects of Theories of Measurement', *Journal of Symbolic Logic* **23** (1958) 113–28.

Slutsky, E., 'Sulla Teoria del Bilancio del Consomatore', *Giornale degli Economist* **51** (1915) 1–26.

Stigler, G. J., 'The Development of Utility Theory', *Journal of Political Economy* **58** (1950) 307–27 and 373–96.

Strotz, R. H., 'Cardinal Utility', *American Economic Review* **43** (1953) 384–97.

Suppes, P. and Winet, M., 'An Axiomatization of Utility Based on the Notion of Utility Differences', *Management Science* **1** (1955) 259–70.

Suppes, P. and J. L. Zinnes, 'Basic Measurement Theory', in *Handbook of Mathematical Psychology*, Vol. 1 (ed. by R. D. Luce, R. R. Bush, and E. Galanter), John Wiley and Sons, New York, 1963.

von Neumann, J. and Morgenstern, O., *Theory of Games and Economic Behavior*, 2nd ed., Princeton University Press, Princeton, New Jersey, 1947.

Weldon, J. C., 'A Note on Measures of Utility', *Canadian Journal of Economics and Political Science* **16** (1950) 227–33.

I. J. GOOD

INFORMATION, REWARDS, AND QUASI-UTILITIES*

ABSTRACT. A set of mutually exclusive and exhaustive outcomes are assumed to have in some sense 'true' probabilities p_i ($i = 1, 2, \ldots n$) and a forecaster estimates these probabilities as q_i. After the outcome is known, the forecaster's client pays him a fee $f_i(q_1, \ldots, q_n; \pi_1, \ldots, \pi_n) = f_i(\mathbf{q}; \pi)$ where π denotes the client's estimates of the probabilities. If this fee has the property that its 'true' expectation is maximized when $\mathbf{q} = \mathbf{p}$ it is called an 'accuracy incentive'. (When $\mathbf{p}$ denotes the forecaster's true attainable subjective probabilities, the fee is also an 'honesty incentive'.) A brief historical survey of the problem of finding accuracy incentives is given, together with a recommendation that a forecaster should not be the same person as a valuer of the outcomes. One form of accuracy incentive (which depends on a parameter $\beta \geqslant 1$) is analogous to a generalized surprise index. It has the merit of being 'splitative', that is, it is unaffected if an outcome is arbitrarily split up into more than one outcome. It is conjectured that there are no other splitative accuracy incentives, but β remains to be chosen. When $\beta = 1$ the fee has an additive property that seems fairly desirable.

All accuracy incentives are of the form grad $g(\mathbf{q})$ where $g(\mathbf{q})$ satisfies certain conditions and is named a 'McCarthy potential'. For choosing between different (splitative) potentials it is proposed that the 'relative peakedness' (div grad $g)/g$, or rather some average of it, be used as a criterion.

A relationship is given to some recent work on the problem of measuring the utility of a distribution to its user. Since the knowledge to which a distribution is to be put is often vague it can be advisable to make use of quasi-utilities, where a quasi-utility is any target function that we try to maximize, preferably having an appropriate additive property.

A producer of information can be regarded as an estimator of probabilities or a *forecaster*, whether he is a statistician, a meteorologist, a medical diagnostician, a stockbroker, a racing tipster, or merely an inanimate communication channel. I am going to consider the problem of how the forecaster's client might reward the forecaster in such a manner as to encourage him to make accurate estimates of probabilities. (The rewarding of an inanimate communication channel can be interpreted in terms of a score for choosing between two or more channels by means of cumulative scores over a fairly long run.) In broad outline, the method is to let the forecaster make his estimates of the relevant probabilities and then to pay him a fee that depends both on these estimates and on which of the possible events happens later. As far as I know this problem was first considered by Brier (1950) and independently in a section of my paper on rational decisions (Good, 1952), but I gave the section the inadequate heading 'Fair fees'. McCarthy (1956) used the description 'a payoff rule that keeps the forecaster honest', which is a much better name. Here I shall call it an *incentive for accurate probability estimation*, or shortly an *accuracy incentive*. A more complete definition is given later.

Let us begin with an example. Suppose that an inventor offers to sell an idea to a firm. The firm might ask the advice of an expert whether to buy the idea. If the expert believes that the idea would be very valuable if it is successful, but that the probability of success is small, he might recommend against the purchase in order to protect his *own* reputation. In such circumstances it would be wise for the firm to hire *two* experts, one to estimate the probability and one to estimate the expected value if successful, a forecaster and a valuer (Good, 1968a), although this might be oversimplified because there might be various degrees of success. In some circumstances the forecasting and the valuing might anyway involve expertise of two different kinds; in fact the client might be better able than the forecaster to judge the utilities. Then again, if the forecaster is a meteorologist he might not even know what application the client has in mind for the forecasts, and so would be in the dark regarding the utility of his predictions. Meteorology is a good example of our problem because the outcomes are observable after little delay. It was the first example considered but even now in many countries weather forecasts are given without specification of the probabilities so that the user has to guess the meanings of the forecasts. (See Brier, 1950; Good, 1952, p. 112; 1954/57, p. 31 or 33, *Cambridge Daily News*, September 25, 1951, p. 10; McCarthy, 1956; Winkler and Murphy, 1968, and references in this latter paper.)

Having separated the forecaster from the valuer we now have to determine the accuracy incentive for the forecaster.

Let us suppose that there are n possible mutually exclusive outcomes and that these have credibilities $p_1, p_2, ..., p_n$ on all evidence that it is practicable to make available. (Admittedly this is not a clear-cut definition and a more complete analysis would make allowance for the costs of searching for information.) These credibilities can be regarded as the personal or subjective probabilities of a *demiurge*. A demiurge can be regarded as a perfectly rational man who also has perfect judgment about where to obtain information, and also allows for practicability in obtaining it. Since he allows for practicability he is not an omnipotent god. We assume that the credibilities are unknown to the client but that he has his own subjective initial (prior) probabilities $\pi_1, \pi_2, ..., \pi_n$. A forecaster now, with or without experimentation, provides his estimates $q_1, q_2, ..., q_n$ of the probabilities ($\sum q_i = \sum p_i = \sum \pi_i = 1$). These estimates are to be made subject to a time limit. The forecaster is given a down payment to cover his expenses and the value of his time, and in addition the client contracts to pay an accuracy-incentive fee which will depend on the event that occurs. This fee cannot depend mathematically on the p's because the client never knows these, but it is to be a function of the π's and q's. [Brier (1950) did not allow for the π's and Good (1952) did so only exiguously. They were brought in explicitly by Good (1954/57.] The forecaster should be told the π's in advance.

One method for selecting the accuracy-incentive fee is to imagine that, by means of an incantation, the demiurge is manifested and that he enters into competition with the human forecaster, but that the client is unable to distinguish the human from the demiurge. The client then considers the hypothesis H that the forecaster is the demiurge. If the ith event occurs this provides weight of evidence $\log(q_i/p_i)$ in favour of H. Weight of evidence is defined as the logarithm of the ratio by which the *odds* of a hypothesis are multiplied in virtue of the evidence: See, for example, Peirce (1878), Good (1950, 1968). The unit of measurement depends on the base of logarithms. Turing (private communication, 1940) invented the name 'ban' for when the base is 10, and 'deciban' for one tenth of this unit, by analogy with the decibel in acoustics.

If the client knew the p's he could not do better than accumulate these weights of evidence on a number of statistically independent occasions in order to identify the demiurge. Again, if there were yet another (human) forecaster, whose probability estimates were $q'_1, q'_2, ..., q'_n$, and if he could be paid in accordance with the analogous formula ($\log(q'_i/p_i)$), the difference between the payments to the two forecasters would be $\log(q_i/q'_i)$. This difference would then be a natural measure of how much more like a demiurge the first forecaster was than the second, on the evidence available, and it has the advantage that it does not depend on the p's. The feature of distinguishing the forecasters would be equally well achieved if they were separately paid $a_i + \log q_i$ and $a_i + \log q'_i$, where a_i is any function of i. But if an expert's probability for the ith event is π_i, then he has not caused the client to change his own estimate of the probability, and so it seems reasonable to take $a_i = -\log \pi_i$. This intuitive argument suggests that an appropriate fee might be $\log(q_i/\pi_i)$ when the ith event occurs (Good, 1954/57). This fee can be interpreted as the weight of evidence in favour of the hypothesis that the forecaster is the demiurge rather than the client! It is of course positive if $q_i > \pi_i$ and negative if $q_i < \pi_i$ when the base of logarithms exceeds unity, as we shall always assume. In fact we might as well take the base of logarithms as e for theoretical convenience since we have not specified the unit in terms of which the payment is to be made.

The need to introduce some probability other than q_i into the formula can be seen by thinking of the corresponding continuous problem in which a probability density function is estimated. The probability of a precise outcome is zero in a continuous model, and we cannot use the logarithm of zero. If instead we use the logarithm of the probability density the result would not be invariant under a transformation of the independent variable x. In order to attain invariance it is natural instead to consider the logarithm of the ratio of two probability densities. The obvious ones to use are the density $q(x)$ as estimated by the forecaster and the initial density $\pi(x)$ as estimated by the client. The suggested fee is then $\log(q(x)/\pi(x))$. If we use this formula for the continuous case we can scarcely avoid using $\log(q_i/\pi_i)$ in the discrete case, because the continuous case can be regarded as discrete if the variable is measured to exactly a billion places of decimals.

In the opinion of the demiurge, the expectation of the fee, apart from the down payment, is

$$\sum_i p_i \log(q_i/\pi_i) \quad \text{or} \quad \int p(x) \log \frac{q(x)}{\pi(x)} \, dx$$

depending on whether we are talking about the discrete or continuous problem. (These expressions are called 'trientropies' by Good (1969). Of course $p(x)$ denotes the probability density as estimated by the demiurge.) Less clearly this can be described as the 'objective' expectation. For variations in the q's it is maximized by taking the q's equal to the p's and *this is the defining property of an accuracy incentive.* (The maximum is of the form of an expected weight of evidence or dientropy when the above-mentioned logarithmic pay-offs are used.) Clearly an accuracy incentive can be multiplied by an arbitrary constant without losing its defining property: in the present example it is like giving a financial value to a deciban. An accuracy incentive is called a 'proper' fee by Winkler and Murphy (1968), except that they have in mind the true subjective probabilities of the forecaster rather than those of a demiurge. Hence their emphasis is on honesty, as was McCarthy's, rather than on 'accuracy'. But the mathematical problems and most of the philosophy and applications are essentially the same under the two interpretations.

The fee as defined above, including the down payment, is additive in the sense that if the same method of payment is used on two or more completely independent occasions, we can if we like regard the whole set of occasions as if it were a single occasion, provided of course that the down payments are added, as is reasonable. This assumes however that the deciban is to be taken as the same number of dollars or rubles on each occasion, and this is not an obvious requirement. But is we merely assume that a deciban is worth a constant amount *within* each application, then additivity would be required after dividing the individual fees by these scale factors. In this weaker sense additivity does seem to be a very reasonable requirement.

Closely related to the invariance of the formula for the continuous model is a property of the discrete formula which I shall call the splitting property or *splitativity*. Imagine that the ith event is arbitrarily split into two events by combining it with the outcome of a randomizing device whose behaviour is fully known to the forecaster and to his client. This ought to make no difference to the fee. The fee $\log(q_i/\pi_i)$ has the splitting property because, whatever the outcome of the randomizing device, both q_i and π_i are multiplied by the same factor. Splitativity is a more compelling desideratum than additivity especially for the continuous model. For in the continuous model we can always fully meaningfully split any given 'event' into smaller events, where an event refers to an interval of values of the variable.

Let us now consider what other accuracy-incentive fees might be entertained. If the fee when the ith event occurs is $f(q_i)$ then we require that $\sum_i p_i f(q_i)$ should be maximized by taking the q's equal to the p's. McCarthy says that Gleason (unpublished) showed that $f(q)$ must be of the form $a + b \log q$. This result is correct when $n \geqslant 3$ but not when $n = 2$, and a counter-example due to Martin Beckman is mentioned by Marschak (1959,

p. 96). (There is an analogous breakdown of a theorem on probability estimation due to William Ernest Johnson and R. B. Braithwaite; see Good (1965, p. 26). The reason for the breakdown is similar. The trouble is that the Lagrange condition tells us that $p_i f'(p_i)$ is mathematically independent of i, and for $n = 2$ this states merely that $(1+y) f'((1+y)/2)$ is an even function of y.)

Next suppose that the accuracy incentive is allowed to be of the form $f_i(q_i)$. Then, when $n \geq 3$, we can readily see that f_i must be of the form $f_i(q_i) = a_i + \log q_i$. This is the formula proposed earlier for intuitive reasons. The question arises whether some other proof might be found that would cover the case $n = 2$. In fact a proof can be obtained if the splitting property is assumed. If f_i is a function of q_i and π_i alone; then the splitting property shows that f_i is a function of q_i/π_i alone. The accuracy-incentive property then shows that $p_i f'_i(p_i/\pi_i)$ is independent of i and the logarithmic formula $b_i + \log(q_i/\pi_i)$ then follows if $n \geq 3$, where b_i does not depend on the q's or the π's. The constant b_i has no effect on the incentive property of the payoff and it should probably be taken as zero for the reasons given before. When $n = 2$, we have

$$\frac{p}{\pi} f'\left(\frac{p}{\pi}\right) = \frac{1-p}{1-\pi} f'\left(\frac{1-p}{1-\pi}\right)$$

and since this is true for all π we again obtain the logarithmic formula.

McCarthy (1956) considered pay-off functions of the more general form $f_i(\mathbf{q})$ where $\mathbf{q}$ denotes $(q_1, ..., q_n)$. (The scalar q and the 'vector' $\mathbf{q}$ should not be confused.) He stated the following theorem without proof: the payoff function is an accuracy incentive if and only if $\mathbf{f}(\mathbf{q})$ is the gradient of $g(\mathbf{q})$, by which I mean

$$f_i(q_1, q_2, ..., q_n) = \frac{\partial}{\partial q_i} g(q_1, q_2, ..., q_n),$$

where $g(\mathbf{q})$ is a function convex from below, homogeneous and of the first degree. We could describe g as a convex potential function. Moreover, if the forecaster maximizes his expectation by taking $\mathbf{q} = \mathbf{p}$, then his expectation is equal to $g(\mathbf{p})$, as follows from Euler's theorem on homogeneous functions.

Marschak (1959) stated that McCarthy's theorem is incomplete in that the logarithmic pay-off, $\log q_i$, can be derived from a convex function but not from one that is homogeneous. But A. D. Hendrickson (private communication) has noticed that every convex function g defined on the space for which $p_i \geq 0$ ($i = 1, 2, ..., n$), $\sum p_i = 1$ can be extended to the whole space for which the p's are non-negative by writing $g(\lambda p) = \lambda g(p)$ ($\lambda > 0$), and the extended function will still be convex. For example, the function $\sum p_i \log p_i$ can be written $\sum p_i \log(p_i / \sum p_j)$. Also he has recently proved (unpublished) that McCarthy's result is correct if the pay-off functions are continuous in the *closed* region $\sum q_i = 1$, whereas the logarithmic functions are continuous only in the open region.[1] An example of an accuracy-incentive that is continuous in the closed region is the quadratic incentive (Brier, 1950; Winkler and Murphy, 1968),

$$2q_i - 1 - \sum_j q_j^2$$

for which the McCarthy potential function, written in the form of a homogeneous function of degree 1, is

$$(\sum p_j^2 / \sum p_j) - \sum p_j.$$

Another example of a continuous accuracy incentive is (Good, 1970):

$$h_i(q; \pi; \beta) = \frac{1}{\beta - 1} \left\{ \frac{(q_i/\pi_i)^{\beta-1}}{[\sum q_j^\beta / \pi_j^{\beta-1}]^{1-1/\beta}} - 1 \right\} \quad (\beta > 1).$$

(This is analogous to the generalization of my generalization of the surprise index: Weaver (1948), Good (1954/57, 1956), Rényi (1961).) This is splitative and *therefore* its continuous form

$$h(x) = h(x; q; \pi; \beta) = \frac{1}{\beta - 1} \left\{ \frac{q(x)/\pi(x)}{[\int q^\beta / \pi^{\beta-1} \, dx]^{1/\beta}} \right\}^{\beta-1} - \frac{1}{\beta - 1}$$

is invariant with respect to transformations of the x coordinate. (The theorem implicit here is self-evident but quite interesting.) When $\beta \to 1 + 0$, $h_i \to \log(q_i/\pi_i)$ and $h(x) \to \log(q(x)/\pi(x))$. This suggests the conjecture that every accuracy-incentive function is either the gradient of a McCarthy potential or the limiting form of one. Among these I conjecture further that the only splitative accuracy-incentive functions are linear combinations of a finite or infinite number (including a continuous infinity) of h_i's for various values of β (including $\beta = 1 + 0$). [Among these the only additive one is $\log(q_i/\pi_i)$.] As support for the latter conjecture, note that any sum (or integral) of expressions of the form

$$\phi\left[\frac{q_i}{\pi_i}, \sum_j q_j \chi\left(\frac{q_j}{\pi_j}\right)\right]$$

is splitative and this seems to be the most general such form. If it is now assumed that each such expression satisfies McCarthy's theorem then it follows that the only splitative accuracy-incentive functions are linear combinations of the h_i's for various values of β. A proof of this can be based on the irrotational property of the gradient of a potential, which property leads to a soluble functional equation, but unfortunately the details are somewhat complicated and will be omitted.

I. CHOOSING BETWEEN (SPLITATIVE) ACCURACY INCENTIVES

Whether or not the above conjectures are correct the question arises how to choose between the various (splitative) accuracy incentives. The client would like to *maximize* the forecaster's incentive in some sense, but without increasing the 'expected' fee, where the expectation requires definition. A way of interpreting this desideratum is to demand that the 'peakedness' of the surface $z = f_i(\mathbf{q})$ at the point $\mathbf{q} = \mathbf{p}$ should be maximized over all accuracy-incentive fees $f_1, \ldots, f_n$, at least in some average sense (averaged for all values of $\mathbf{p}$). When $n = 2$ the peakedness could again be reasonably interpreted either as kurtosis or as curvature, and for $n \geqslant 3$ either as something analogous to Gaussian or to mean curvature or as some generalization of kurtosis. Among the splitative pay-off functions h_i, these various definitions would in principle lead to an 'optimal' specification of the parameter β. Perhaps one of the definitions would support the value $\beta = = 1 + 0$ corresponding to the logarithmic pay-off function. (Presumably it is the only additive accuracy incentive.)

A promising approach is to expand the expectation

$$\sum_i p_i f_i(\mathbf{p} + \boldsymbol{\varepsilon}),$$

where $\boldsymbol{\varepsilon}$ is a small vector, by Taylor's theorem in several variables. The terms of first order vanish because the expectation is maximized at $\boldsymbol{\varepsilon} = 0$.

If we denote derivatives by subscripts, so that f_i is the same function as g_i, we obtain

$$g(\mathbf{p}) + \frac{1}{2!}\sum_{ijk} \varepsilon_j \varepsilon_k p_i g_{ijk}(\mathbf{p}) + \frac{1}{3!}\sum_{ijkl} \varepsilon_j \varepsilon_k \varepsilon_l p_i g_{ijkl}(\mathbf{p}) + \cdots,$$

where we have used Euler's theorem $\sum p_i g_i(\mathbf{p}) = g(\mathbf{p})$. Now $g_{jk}(p)$, $g_{jhl}(p), \ldots$ are homogeneous and of degrees $-1, -2, \ldots$, so, again by Euler's theorem,

$$\sum_i p_i g_{ijk}(\mathbf{p}) = -g_{jk}(\mathbf{p}), \quad \sum_i p_i g_{ijkl}(\mathbf{p}) = -2g_{jkl}(p), \ldots.$$

Therefore

$$\sum_i p_i g_i(\mathbf{p} + \boldsymbol{\varepsilon}) = g(\mathbf{p}) - \frac{1}{2!}\sum_{jk} \varepsilon_j \varepsilon_k g_{jk}(\mathbf{p}) -$$

$$- \frac{2}{3!}\sum_{jkl} \varepsilon_j \varepsilon_k \varepsilon_l g_{jkl}(\mathbf{p}) - \cdots.$$

If we take into account terms up to the second order, a natural measure for peakedness, analogous to the mean curvature of a surface, is the trace of the matrix $\{g_{ij}(\mathbf{p})/g(\mathbf{p})\}$, that is,

$$\frac{1}{g(\mathbf{p})}\sum_j \frac{\partial^2 g}{\partial p_j^2} = \frac{\text{div grad } g}{g} = \frac{\nabla^2 g}{g}.$$

We are thus presented with the following interesting mathematical problem: Maximize $(\nabla^2 g)/g$, where g is continuous, convex from below, and is homogeneous of degree 1, or is the limit of a sequence of such functions. Consider especially the 'splitative' potentials. It might be necessary to average $(\nabla^2 g)/g$ in some way in order that the problem should have a solution. In default of a solution to the problem in this general form, at least we could use the criterion $(\nabla^2 g)/g$ to help us to choose between any two specified functions g.

II. RELATIONSHIP TO THE 'UTILITY OF A DISTRIBUTION'

A classical method for encouraging loyalty in an employee is to pay him a certain proportion of the profits (which might be negative). There is no obvious reason why such a scheme would very efficiently encourage accuracy in probability estimation. But suggestions have been made for estimating the utility of a probability distribution (for example, Good 1968c, 1969) and one would expect some relationship of that work to accuracy incentives. Here we shall mention just the bare bones of the other work.

Suppose we wish to estimate the utility of the assertion that the distribution of a random vector $\mathbf{x}$ is Q when, in the opinion of the demiurge, it is really P. (Q and P are of course functions of $\mathbf{x}$. We are first discussing a continuous variable $\mathbf{x}$.) We assume that the use to be made of the estimated distribution Q is optimal, but that this use is unspecified. This assumption is reasonable for the forecaster who is not also a valuer. The theory applies also when Q is to be used for a search such as for a lost submarine, and in this case $\mathbf{x}$ is a two-dimensional or three-dimensional vector. For non-optimal use we cannot assume axiom A5 below. Non-optimal use is also discussed by Good (1969).

The utility depends on many things and can be very difficult to estimate. It is therefore convenient to introduce 'quasi-utilities' where a quasi-utility satisfies various reasonable desiderata. We denote such a quasi-utility by $U(Q|P)$ and suppose that it obeys the following axioms:

A1. $U(Q | H_\mathbf{x})$ is a generalized expectation of $v(\mathbf{x}, \mathbf{y})$, defined as the utility of asserting that the (mathematical) value of the random vector is $\mathbf{y}$ when it is really $\mathbf{x}$. Here $H_\mathbf{x}$ is the Heaviside function that, as a distribution, assigns a definite value to $\mathbf{x}$. By a 'generalized expectation' of v we mean an expression of the form $\phi^{-1} \int \phi(v(\mathbf{x},\mathbf{y})) \, dQ(\mathbf{y})$.

A2. If a constant is added to v then the same constant is added to $U(Q | P)$.

A3. The quasi-utilities for two entirely independent problems are additive, that is $U(QQ^* | PP^*) = U(Q | P) + U(Q^* | P^*)$.

A4. $U(Q | P)$ is invariant under a non-singular transformation of $\mathbf{x}$, $\mathbf{x} = \psi(\mathbf{x}')$, $\mathbf{y} = \psi(\mathbf{y}')$, where the transformed form of v is $v((\psi(\mathbf{x}'), \psi(\mathbf{y}'))$.

A5. $U(P|P) \geq U(Q|P)$.

I conjecture that these axioms imply the following form for U:

$$U(Q | P) = \int \log \{q(\mathbf{x}) | \Delta(\mathbf{x})|^{-1/2}\} \, dP(\mathbf{x})$$

when Q has a density function q, and

$$\Delta(\mathbf{x}) = \left\{ -\left.\frac{\partial^2 v(\mathbf{x}, \mathbf{y})}{\partial y_i \partial y_j}\right|_{\mathbf{y}=\mathbf{x}} \right\}.$$

If the demiurge submitted a report the client would pay the forecaster an amount proportional to $U(Q \mid P)$. Since the demiurge knows, but will not tell, the most natural fee is proportional to $\log\{q(\mathbf{x}) \mid \Delta(\mathbf{x}) \mid^{-1/2}\}$, when $\mathbf{x}$ is the value that occurs, since this is the only function known to have the correct expectation. If $v(\mathbf{x}, \mathbf{y})$ is any twice differentiable function of $(\mathbf{x} - \mathbf{y})^2$, then $\Delta(\mathbf{x})$ reduces to a constant and so drops out of account. Quadratic loss functions are unreasonable when $\mathbf{y}$ is not close to $\mathbf{x}$, but loss functions of the form $A - A \exp\{-B(\mathbf{x}-\mathbf{y})^2\}$ are reasonable when we are dealing with problems of estimation. When we are dealing with mixed problems of estimation and significance a reasonable loss function is

$$A\{1 - e^{-B(\mathbf{y}-\mathbf{x})^2}\} + \frac{C}{\sigma\sqrt{2\pi}} \exp\left\{-\frac{(\mathbf{x}-\xi)^2}{2\sigma^2}\right\} \{1 - e^{-D(\mathbf{y}-\xi)^2}\},$$

where σ is small, D is large, and the null hypothesis asserts $\mathbf{x} = \xi$. (All this can be generalized to arbitrary positive quadratic forms in the place of the squares.) Then $\Delta(\mathbf{x})$ does not drop out of account.

The quasi-utility gained by shifting from an asserted distribution of density $\pi(\mathbf{x})$ to another one, of density $q(\mathbf{x})$, is exactly equal to the expectation in the opinion of the demiurge, of the logarithmic fee $\log(q(\mathbf{x})/\pi(\mathbf{x}))$. Thus there is some relationship between the quasi-utility formula and the logarithmic accuracy-incentive fee.

III. ENCOURAGEMENT OF RESEARCH

Dr. Gordon Tullock (private communication, 1970) has raised the question of how to encourage research in probabilistic forecasting without encouraging incompetent research. (He mentioned the example of stating a probability density for next year's Gross National Product.) The problem differs from that considered in the present paper in that there could then be a *group* of forecasters among whom must be shared a certain pool of money. Perhaps an adequate solution would be to allot shares proportional to accuracy-incentive fees, where the less good forecasts would attract negative rewards. The good forecasts would be subsidized by the bad ones

As Dr. Tullock remarks, there is much to be said for subsidizing research *after* it is done instead of by 'projects'.

IV. UNSOLVED PROBLEMS

The present paper contains a few mathematical conjectures whose proofs I have so far been unable to supply. I suspect that the proofs or counter-examples will not be easy to find.

An aspect I have slurred over is the cost to the *forecaster* of obtaining information: in practice there is a whole hierarchy and even a network of forecasters, forming an information market. Perhaps the way to deal with this complication is simply to add the forecaster's expenses to the accuracy incentive fee. These expenses would include the cost of experimentation, of computation, and of acquiring information in other ways, and some limit would have to be set on the total expenses.

The so-called 'imperfections' of the ordinary market are due to the cost of information and when computers bring these costs down, the economic repercussions will be enormous under both capitalism and socialism, in fact both systems might become possible for the first time.

Virginia Polytechnic Institute and State University, Blacksburg

BIBLIOGRAPHY

Brier, G. W., 'Verification of Forecasts Expressed in Terms of Probability', *Mon. Wea. Rev.* **78** (1950), 1–3.

Good, I. J., *Probability and the Weighing of Evidence*, Charles Griffin, London, Hafners, New York, 1950, pp. 119.

Good, I. J., 'Rational Decisions', *JRSS B* **14** (1952), 107–14.

Good, I. J., 'Mathematical Tools', Chapter 3 of *Uncertainty and Business Decisions*, Liverpool, 2nd ed. 1954/57, pp. 20–36; based on a symposium in the Economics section of the British Association, 1953.

Good, I. J., 'The Surprise Index for the Multivariate Normal Distribution', *Annals Math. Statist.* **27** (1956), 1130–5; corrections, *l.c.* **28** (1957).

Good, I. J., *The Estimation of Probabilities: An Essay on Modern Bayesian Methods*. M.I.T. Press, 1965, pp. xii + 109.

Good, I. J., 'Some Statistical Methods in Machine-Intelligence Research', *Virginia J. Sc.* **19** (1968), 101–10; reprinted with improvements in *Math. Biosc.* **6** (1970), 185–208.

Good, I. J., 'Corroboration, Explanation, Evolving Probability, Simplicity, and a Sharpened Razor', *Brit. J. Philos. Sc.* **19** (1968b), 123–43.

Good, I. J., 'Utility of a Distribution', *Nature* **219** (1968c), 1392.

Good, I. J., 'What is the Use of a Distribution?', *Multivariate Analysis – II* (ed. by P. R. Krishnaiah), Academic Press, New York, 1969, pp. 183–203.

Good, I. J., 'Contribution to the Discussion of a Paper by R. J. Buehler, in *Proceedings of the International Statistics Symposium (at Waterloo, Ontario)* Holt, Rinehart and Winston of Canada, Toronto, 1970.

Marschak, J. (1959), 'Remarks on the Economics of Information', in *Contributions to Scientific Research in Management*, University of California, Berkeley, pp. 79–98.

McCarthy, John, 'Measures of the Value of Information', *Proc. Nat. Acad. Sc.* **42** (1956), 654–5.

Peirce, Charles Saunders, 'The Probability of Induction', *Popular Science Monthly* (1878), reprinted in *The World of Mathematics*, **2** (ed. by J. R. Newman), Simon and Schuster, New York, 1956, 1341–54.

Rényi, A., 'On Measures of Entropy and Information', in *Proc. Fourth Berkeley Symp. Math. Statist. Prob.* (ed. by J. Neyman; Univ. of Calif. Press, Berkeley, 1961, pp. 547–61).

Weaver, Warren, 'Probability, Rarity, Interest and Surprise', *Scientific Monthly* **67**, (1948), 390–2.

Winkler, R. L. and Murphy, Allan H., '"Good" Probability Assessors', *J. Appl. Meteorology* **7** (1968), 751–8.

NOTES

* Revision of an invited paper presented at the Second World Congress of the Econometric Society, Cambridge, England, September 1970.

[1] I understand he has now generalized the result to include the logarithmic case.

HECTOR-NERI CASTAÑEDA

OPEN ACTION, UTILITY, AND UTILITARIANISM*

(i) ... when the question is which among several courses still open to a man he *ought* to choose.
G. E. Moore, *Ethics*, Chapter 5

(ii) ... an obligation must be an obligation, not to *do* something, but to perform an activity of a totally different kind, that of setting or exerting ourselves to do something, i.e., to bring something about.
H. A. Prichard, 'Duty and Ignorance of Fact'

This essay is another installment in a continuing examination of the logical foundations of utilitarianism.[1] Here we discuss act-utilitarianism.

Utilitarianism is a family of views that attempt to articulate *in concreto* the ideal that one ought always to do the best. For instance, according to simple act-utilitarianisms an agent X ought always to choose to do, and to do, an action A if and only if X's doing A causes at least as great a value of a certain kind as any other alternative action open to X in his circumstances. We call these views 'simple' because they make the value-maximizing condition (i.e., causing at least as great a value) both sufficient and necessary for obligatoriness. The moral criticisms of such views are well known, and do not concern us here. They cannot eliminate act-utilitarianism as a source of prima facie duties or as a criterion for the solution of some conflicts of duties. What concerns us here is the logical foundation of, indeed, the very intelligibility of, act-utilitarianism. Palpably, a utilitarianism, for that matter, any theory of obligation, is intelligible to the extent that there is an understanding of what an action in general is[2], and in particular to the extent that there is an understanding of the concepts *alternative action, open action*, and *best consequence*.[3] These three concepts concern us here. Their analysis is crucial for the very formulation of act-utilitarianism. In fact, elsewhere,[4] where we began the study of the logical foundations of utilitarianism, we have already argued that the value-maximizing condition cannot be necessary for obligatoriness, i.e., that a viable act-utilitarianism cannot be simple.

But our main purpose here is not critical, except for one foundational point. The simple utilitarianisms above presented cannot provide guidance in the most typical cases of action and choice, even though most forms of act-utilitarianism are realistic theories of obligation; they are concerned, not with what one ought ideally and unfeasibly to do, but with the question: Which among the courses of action *open* to an agent ought he to choose? Realism requires that in this question the chosen course of action be not merely an action that the agent ought to perform, but an action that the agent ought to perform or just to *try* to perform: For sometimes it is just not open to the agent to do an action that would cause the greatest relevant good, but it is only open to him just to *try* to do one such action. The problem here is that the criterion for choosing what one is to try to do cannot be causing the greatest good. Clearly, for trying, probabilities matter; yet (as will be shown below) how to combine probabilities and values is not an easy task.

Our purpose here is constructive. In Section I, we offer a contribution to the analysis of *open alternative action*, distinguish between a set of actions for performance and a set of actions for choice, and structure the relationships among the duties to choose, to do, and to try to do. In Section II, from the vantage position gained in Section I, we examine some ways of meeting the serious problem of combining value and probability that have already been explored in utility theory and the modern theory of decision under risk. Since the ways in question are not satisfactory, we conclude the essay with a very tentative proposal for combining probabilities and value in a version of act-utilitarianism. With this proposal, mounted on the analysis of open alternative action, we have here at least a coherent formulation of act-utilitarianism. But it is not a simple utilitarianism.

I. OPEN ALTERNATIVE ACTIONS

1. *Performances Alternative Actions*

Consider a man, to be called *Utilius*, who is locked up in a room and knows that a mechanism has been turned on which will explode a bomb in the midst of a political gathering. Fortunately, before Utilius there is the only thing in the room, to wit, an electric gadget with three buttons:

(i) button b_1 which if depressed, will just turn off the mechanism operating the bomb;

(ii) button b_2 which if depressed, will turn on a mechanism that will *either* just stop the mechanism operating the bomb *or* both stop the bomb and turn on a fire alarm;

(iii) button b_3 that produces the same effects as button b_2.

Thus, Utilius has to make a choice among three alternatives:

1. $S \sim R$: stopping the bomb without ringing the fire alarm;
2. $S \sim R \vee SR$: either stopping the bomb without ringing the fire alarm or both stopping the bomb and ringing the alarm;
3. $\sim S \sim R$: doing nothing to stop the bomb (or ring the alarm).

Of course, Utilius also has to choose among:

1'. depressing button b_1;
2'. depressing button b_2 or b_3;
  $2'_a$: depressing button b_2; 2_n: depressing button b_3;
3'. depressing no button at all.

Undoubtedly, there are other sets of actions from which he may very well have to make further choices, e.g., whether to depress the button he wishes to depress with his left or with his right hand, whether to depress standing or kneeling. But Utilius's primary consideration is his bomb issue. With regard to this issue the primary relevant actions are *1, 2,* and *3* above. Actions *1', 2',* and *3'* are relevant to that issue only as means of bringing about actions *1, 2,* and *3*. All other actions he can perform in that room are either wholly irrelevant to his burning issue or are derivatively relevant as means to the means *1'*, or *2'* or *3'* or as means to more remote links in the chains of means for bringing about *1* or *2* or *3*.

With respect ot the bomb issue, the triple of actions *(1,2,3)*, i.e., $(S \sim R, S \sim R \vee SR, \sim S \sim R)$, is exhaustive of the alternative courses of action available to Utilius in his circumstances. Evidently, he is in no position to perform $\sim SR$, i.e., to ring the alarm without stopping the bomb, let alone to choose rationally, truly believing himself to be able to perform $\sim SR$. Furthermore, in terms of the set-up surrounding Utilius the members of $(S \sim R, S \sim R \vee SR, \sim S \sim R)$ are pairwise *choice*-incompatible. Utilius must *choose* between $S \sim R$ and $S \sim R \vee SR$. On the contrary, the sets *(1', 2'a, 3')* and *(1, 2'b, 3')*, each of which represents a complete list of Utilius's means of doing something about his bomb issue, are not exhaustive with respect to button-depressing: each leaves another button available. The set ($1'$, $2'$, $3'$) is exhaustive with respect to button-depressing. But it can fail to be pairwise incompatible. It may be the case, for instance, that Utilius can depress both buttons b_1 and b_2 simultaneously. Naturally, if the wiring diagram of the gadget conforms to logical relations, depressing button b_1 and b_2 simultaneously has exactly the same output as depressing button b_1 alone, for $(S \sim R)$ & $(S \sim R \vee SR)$ is equivalent to $S \sim R$. Yet, Utilius's doing one action about the bomb in the set (1, 2, 3) is identical, although only contingently, with his doing the corresponding primed action in the set ($1'$, $2'$, $3'$). That is, the actions in these two sets are abstractly different, but there is just one event or change in the world that Utilius is going to insert, in the causal structure of his circumstances, for other changes to be wrought out that make him the agent of actions constituted by such changes.

In any circumstances whatever the set of actions $(S \sim R, S \sim R \vee SR, S \sim R)$ is redundant as a *performance set*. If Utilius performs $S \sim R$, regardless of how he does it, he automatically performs $S \sim R \vee SR$. Furthermore, if he performs $S \sim R \vee SR$ he either performs $S \sim R$ or he performs SR. If he depresses button b_2 and the bomb is stopped while the fire alarm rings, Utilius performs SR: he is liable or praiseworthy just as if he had depressed button b_1. Clearly, the non-redundant set of performances he can probably produce is $(S \sim R, SR, \sim S \sim R)$. Yet in the circumstances as described above, the set $(S \sim R, S \sim R \vee SR, S \sim R)$ is not redundant as a *choice set*. First, a mechanism is available to Utilius which allows him to choose $S \sim R \vee SR$ without choosing either disjunct. Second, in his circumstances Utilius cannot effectively choose to bring about his performance of SR. If he wants to perform SR he can only choose $S \sim R \vee SR$ and hope that the second disjunct be actualized.

We must therefore, distinguish: (i) sets of actions as alternative performances of an agent from sets of actions as alternative choices for performance, and (ii) each of these types of sets from sets of actions as means for performances.

2. *The Structure of Action: Action Machines*

In general, abstracting from the irrelevant details and the specific features of Utilius's case, we can see that a situation in which an agent ought to choose what to do is a complex causal structure that includes the following elements:

(A) A dominant purpose or point of view W that organizes a subset of the causal links and possible changes issuing from the agent's simplest actions into a hierarchy of actions;
(B) The set of possible performances by the agent, determined both by the dominant purpose or viewpoint W and the network of causal links that start from the agent's simplest actions: to be called *the agent's output actions with respect to W*;
(C) The set of possible choices available to the agent, determined by the output actions and the causal network: to be called *the agent's input actions with respect to W*;
(D) The set of possible ultimate means available to the agent;
(E) Sets of possible actions which are means to means to the agent's bringing about his performances;
(F) The set of all remaining possible actions, which are wholly irrelevant to the dominant purpose and to subsidiary purposes pertaining to the choice of means.

We shall refer to (D) and (E) as the mechanics of the input and the output actions. The crucial principle here is:

EP. The agent's performing an output action A with respect to viewpoint W is identical with his performing an input action B with respect to W, and the latter is identical with his performing a means action C for B, and this is identical with his performing a lower means action D for C, and so forth down to some simplest basic action, which in the case of contemporary earthlings is a bodily movement.

We are concerned here with (C), the set of input actions with respect to some viewpoint W. Thus we shall simply assume that each agent is at the focal point of a causal network and that this network includes the mechanics of the input and output actions under consideration. We need not differentiate between alternative means for an input action (e.g., depressing button b_2 and depressing button b_3 in Utilius's case above). Furthermore, the nature of the means is of no importance here. It is immaterial for us here, for instance, whether Utilius mails a letter by walking two miles to a mailbox, or whether he mails the letter by asking his son to give it to the mailman at the door, or by pushing a button in a mailing contrivance.

Thus, for convenience we may imagine that our agents act by simply pushing buttons in an *action machine* like the gadget Utilius operated in Subsection 1 above. We shall do so, but with one important change.

It will be recalled that Utilius was in a position to choose the disjunction stopping the bomb without ringing the warning bell, or stopping the bomb and ringing the bell ($S \sim R \vee SR$). We deliberately refrained from injecting there probability considerations. But it is perfectly obvious that button b_2 may be connected to a mechanism that has more chances of activating the mechanism that brings about the first disjunct than of activating a mechanism that brings about the second disjunct. Thus we may suppose that our action machine has, instead of pushbuttons, circular dials, with unit radius, divided up in sectors of different area. Each sector represents a disjunct of a disjunction assigned to the dial, and the area of the sector represents the probability that the mechanism connected to the dial will bring about the disjunct represented by the sector. Thus, the ultimate means action the agent can perform is simply to spin the needle of the dial representing the input action he has chosen.

3. *Open Actions*

The open alternatives are in fact the input actions. That is, a course of action open to an agent in some circumstances C is a course which the agent not only can perform in C, but can also choose effectively to perform in C. Obviously, the sense of 'can' in use here is a very robust contingent one involving both powers and abilities as well as definite causal circumstances. Nevertheless, in spite of the intimate connection between this sense of 'can' and the sense in which an alternative action is open, these senses contrast as follows:

Perf. 1.   'X can perform $A \vee B$ in C' entails 'X can perform A in C or X can perform B in C';

Op. 1.   '$A \vee B$ is open to X in C' does *not* entail 'A is open to X in C, or B is open to X in C.'

For the record, let us list other crucial principles of the logic of openness, vis-à-vis the logic of our strong *can*. For convenience, let us abbreviate as follows:

$Ax = X$ does not perform A in C

$\Diamond(Ax)$ = it is logically possible for X to do A in C
$\diamondsuit(Ax)$ = X can (in our strong sense) do A in C
$o(Ax)$ = it is open for X to do A in C
$\Rightarrow$ = entails.

Then our crucial principles are:

Perf. 2. $Ax \Rightarrow \diamondsuit(Ax)$
Perf. 3. $\diamondsuit(Ax) \Rightarrow \Diamond(Ax)$
Perf. 4. $\diamondsuit(Ax \vee Bx) \Rightarrow \diamondsuit(Ax) \vee \diamondsuit(Bx)$
Perf. 5. If $(Ax \Rightarrow Bx)$, then $\diamondsuit(Ax) \Rightarrow \diamondsuit(Bx)$
Op. 2. $o(Ax) \Rightarrow \diamondsuit(Ax)$
Op. 3. $o(Ax) \Rightarrow o(\sim Ax)$
Op. 4. If $Ax \Rightarrow Bx$, then $o(Ax) \Rightarrow o(Bx)$.

4. *Choice Spaces: Open Alternatives*

We must tackle the problem of determining the set of input actions, i.e., the set of open alternatives from which an agent ought to choose what to do. As said in Subsection 2, this set is determined by both the dominant purpose or viewpoint and the causal network around the agent. We can roughly divide the contribution each makes as follows: (i) the viewpoint determines the content or type of action of the input action; (ii) the causal network determines the openness, and particularly the chances or amounts of probability of each output action with respect to a given input action.

Here we are not analyzing the viewpoint. All that matters here is that the viewpoint determines what is important in the agent's situation by determining the set of output actions and the content of the input actions. Thus, here we may treat the viewpoint simply as the assignment of a set of important or relevant actions to the agent's situation. For reasons that will emerge, we may take the set of relevant actions characterizing a given viewpoint as a set of actions that are units of relevance, i.e., actions which may be logically and ontologically as complex as one desires, but are such that from the viewpoint under consideration only they and compounds of them are relevant, i.e., have the value pertaining to the viewpoint.

We start, then, with a set α of actions that are units of relevance, which we shall call *atomically relevant actions* characteristic of the viewpoint under consideration, and for short *aras*. Let

$$\alpha = \{A_1, ..., A_n\}.$$

Clearly, α determines a set β of actions which are conjunctions of aras or their negations, but not both. There are 2^n conjunctions of the form

$$(A_1)' \,\&\, (A_2)' \,\&\, ... \,\&\, (A_n)', \text{ where each } (A_j)' \text{ is either } A_j \text{ itself}$$
or $(\sim A_j)$.

These conjunctions are the full alternatives that the agent can bring out. *Qua* performances they are the basic outcomes that will issue from the agent's choosing and doing. We shall call them *boas*, as short for *basic output actions* determined by the viewpoint under consideration. Thus:

$$\beta = \{C_1, ..., C_{2^n}\}, \text{ where each } C_i \text{ is a boa}.$$

We may regard the boas as the relevant *axiological atoms*, i.e., as the actions that are immediately assigned value by the viewpoint. This is not required, but it has the advantage of bypassing the question whether the value of a conjunction is the sum of the values of its conjuncts or not. As is well known from elementary logic, all the connective compounds of the aras are either equivalent to a boa or to a disjunction of a boas. Thus, by treating the boas as axiologically basic we reduce the problem of computing values to the problem of reckoning the values of disjunctions given the values of the boas which are their disjuncts.

Since X's performing a boa entails his performing any action entailed by a boa, we consider the actions of the form:

$$(C_1)' \vee (C_2)' \vee ... \vee (C_{2^n})',$$
where each $(C_j)'$ is either C_j or a null disjunct.

We shall call these actions *oas*, as short for *output actions* characteristic of the viewpoint under consideration. Clearly, the boas are oas. When every $(C_j)'$ is null, we find the null action which is, then, also an oa. There are altogether 2^{2^n} oas. Thus, we have the set

$$\gamma = \{D_1, D_2, ..., D_{2^{2^n}}\}, \text{ where each } D_i \text{ is an oa}.$$

Now we bring in the contribution to the set of open alternatives from the causal network surrounding the agent. Consider the action machine. Some of the dials correspond to boas, and some to disjunctive oas. The former simply assign probability 1 to boas; the latter assign probability 1 to disjunctions, and partition this probability into the disjuncts of the disjunctions in question. More precisely, each dial d_j is divided into 2_{2^n} sectors (some of which are null) of area p_i^j in the following correspondence:

$(C_1)', (C_2)', \ldots, (C_2n)'$
$p_1^j, p_2^j, \ldots, p_2^j n,$
where $p_1^j + p_2^j + \cdots + p_2^j n = 1$ and each $p_i^j \geq 0$.

Each of these assignments of a probability distribution to an oa yields a *poa*, i.e., a *probabilitized output action* with respect to the given viewpoint and the agent's causal network.

Since each of the boas is assigned a value, we may consider a poa M_j as a set of triples, thus:

$M_j = \{(C_1, v_1, p_1^j), (C_2, v_2, p_2^j), \ldots, (C_2n, v_2n, p_2^j n)\},$
where v_1 is the value of boa C_i, and p_i^j is the probability assigned by M_j to C_i.

Clearly, there are infinitely many ways in which probability distributions can be assigned to each oa. It is not clear, however, that the causal networks surrounding agents have infinitely many mechanisms that determine, for a given viewpoint, infinitely many poas. Yet, even if nature allows infinitely many mechanisms, it seems safe to assume that acting creatures have a threshold in their perception of probabilities. It seems likely that for the most part, at any rate, an agent need consider just a finite set of poas. Indeed, his causal network may even fail to provide mechanisms for determining a poa for each oa. This was in fact the case of Utilius in Subsection 1 above. In general, continuing to use the model of the action machine, we must note that an agent's action machine at a certain time may have some inoperative dials. Thus, form the whole set of poas that we can form in principle, we must select the poas which are available to the agent, i.e., those poas for which he can perform a means action. These are the *opoas*, or *open poas*. Hence, we arrive at the set

$\omega = \{M_1, M_2, \ldots\},$ where each M_j is an opoa.

The set ω, which may very well be infinite, is, then, the set of open alternatives from which an agent must determine what he ought to choose to do. The set ω is the actual space of choices at time t for an agent X, given a viewpoint and X's causal network at t. And our final analysis of *open alternative for X at time t in situation S with respect to viewpoint W* is simply that an open alternative for X at t in S with respect to W is an open probabilitized output for X at t in S with respect to W, i.e., a probabilitized output action M such that X is in S and X can effectively choose to perform in S at t a means action A with respect to W, and his doing of A would be identical with his doing of M. Also, X's input actions at t with respect to W are his opoas at t with respect to W.

5. *Duties to Choose, to Do, and to Try*

Now we can provide an account of the relationships among what a man ought to do, what he ought to try to do, and what he ought to choose to do, or try to do. First of all, *always what an agent ought to choose to perform is an opoa*. This is true not only with respect to the primary point of view under consideration, but also for the derivative points of view brought in when the agent is choosing the means for his primary opoas. In general, given a set ω of opoas, there is a set of opoas from which the agent ought to choose one. Let us call these *dutiful opoas*, or *dopoas*, and the set of them Δ_ω or simply Δ, when it is clear what the set ω is.

In situation S, then, an agent has a set Δ of dopoas from which he ought to choose what to bring about as an input action for his action machine. He also ought to choose a means action for bringing about the dopoa of his choice.

If $\Delta_\omega = \omega$, then there is normally no action that the agent ought to do or try to do with respect to the point of view that partially determines ω. If Δ has just one member and this is a poa M_j that assigns probability 1 to a given boa C_i, then M_j assigns probability 0 to the other $2^n - 1$ boas. In this case the agent ought to *do* C_i. If for $j = 1, 2 \ldots, k$ and $k > 1$, each opoa M_j in Δ assigns probability 1 to boa C_{ji}, then it is not the case that the agent ought to do any given boa, but he ought to *choose* one of the boas $C_{j_i}, \ldots, C_{j_k}$. In this case, we say that the agent ought to do the disjunction $C_{j_i} \vee \ldots \vee C_{j_k}$. We also say that the agent ought to do a disjunction $C_{j_i} \vee \ldots \vee C_{j_k}$ if the only member of Δ assigns the same probability to each of $C_{j_i}, \ldots, C_{j_k}$ and assigns probability 0 to all the other

2^n-k boas. Finally, but roughly, if an agent ought to do an action A he also ought to do any action B such that his doing A entails his doing B.[5]

Let Δ have some members that assign probability 1 to no boa. Let there be a poa M_h that both assigns probability 1 to a boa C_i and is not in Δ just because M_h is not open to the agent in his situation, i.e., M_h is not in Δ_ω simply because M_h is not in ω. Further, let C_i be the only boa which is assigned by each member of Δ a probability different from 0. Then, the agent's situation is more briefly describable as one in which, although it is not the case that he ought to do C_i, he nevertheless ought to *try* to do C_i. More generally, let D_k be a disjunction $C_i \vee ... \vee C_h$ of boas such that: (i) every member of Δ assigns a nonzero probability to each of $C_i, ..., C_h$; (ii) the poa that assigns probability $1/h$ to each of $C_1, ..., C_h$ would be in Δ if it were in ω (i.e. the poa in question satisfied all conditions for being in Δ except membership in ω), and (iii) D_k is the only disjunction of boas that satisfies (i) and (ii). Then we may briefly say that the agent in question ought, *not* to do D_k, but to *try* to do D_k.

Naturally, the duty to do or try to do some input action M_h carries with it a duty to do some means action A for M_h such that in his situation the agent's doing A is identical with his doing M_h, and, *a fortiori*, with his doing the output action (whatever it may be) that is brought about by A. But what A is and what its own alternatives are cannot be decided merely by an examination of the members of Δ_ω or of the view that determines ω.

6. *The Choice Matrix*

We can represent the structure of an opoa in a simplified way. We order the boas, wholly arbitrarily, or, as we prefer, both by decreasing value and by fiat in case of equal value. Then we order the members of each opoa as follows: the ith triple in an opoa is the one which includes the ith boa. Thus, we need not refer explicitly to the boas in describing an opoa M_j, which is, therefore, an ordered 2^n-tuple $((v_1, p_1^j), ..., (v_{2^n}, p_{2^n}^j))$ of pairs, the members of the latter being a value and a probability, i.e., $0 \leq p_j^i \leq 1$ and $p_1^j + \cdots + p_2^j = 1$.

Now, since each boa is assigned just one and the same value by the viewpoint under consideration, the characteristic of an opoa M_j is simply the probability vector $(p_1^j, ..., p_2^j)$. Thus, the whole set ω or choice space for an agent X in a given situation S at a certain time t can be represented by a probability matrix, to be called the *choice matrix for X in S at t*. For example:

CHART 1

	C_1... 600	C_h... 30	C_i... 100	C_j... 100	C_k... −400	C_{2^n} −800
M_1	0.5	0	0	0	0.5	0
M_2	0	0	0.5	0.5	0	0
M_3	0	1	0	0	0	0
M_4	0	0.4	0.3	0.3	0	0

II. UTILITARIANISM

7. Utilitarianism and Expected Utility

Let us consider Utilius again. He is facing the question which among the several alternatives still open to him he ought to choose. According to act-utilitarianism the answer is very simple: he ought to choose one of the open alternatives that have the highest value. But this answer is nonsensical unless there is a method for computing a value for opoas that assign nonzero probabilities to more than one boa, like M_1, M_2, and M_4 in Chart 1. Inasmuch as the choice from among the opoas is nothing but a choice from probability vectors, it is clear that we are now dealing with an extended, probabilitized sense of value. Nevertheless, it is evident that we cannot simply choose the vector with the highest probability component: for such vector may very well be the one that assigns the highest probability to the worst possible boa.

An opoa is, we say, a 2^n-tuple of pairs of probabilities and values. This is exactly what in utility theory is called a *lottery*, that is, an assignment of probabilities to certain prizes. In our case the prizes are the boas. Now, classical utility theory computes the *expected utility* $u(L)$ of a lottery L with m prizes by means of the formula:

$$u(L) = v_1 p_1 + \cdots + v_m p_m,$$

where v_i is the value of the ith prize.[6]

Thus, connecting utility theory with utilitarianism we can use (1) to determine the value of each of the opoas from which Utilius has to choose. Consider, by way of illustration, the set ω_1 that has just the two members

M_1 and M_2 described in Chart 1. The computations are as follows:

$$u(M_1) = 0.5 \times 600 + 0.5 \times -400 = 100$$
$$u(M_2) = 0.5 \times 100 + 0.5 \times 100 = 100.$$

In this case there is no alternative open to him that Utilius ought to choose, if expected utility is the relevant extension of the value of the boas. Yet in the great lottery of life which morality and utilitarianism mean to provide guidance, we cannot accept expected utility as the proper extension for poas of the values of the boas. It may very well be the case that 400 units of disvalue represent a tremendous catastrophe, which a moral agent ought to refrain from running the risk of bringing about. Indeed, a probability of 0.5 is very large. Even though a moral agent's opoas are lotteries (in the technical sense), the moral agent ought not to be a high-risk gambler. A risk of a catastrophe overrides the prospect of a great bonanza.

In modern utility theory the value of a lottery need not be the lottery's expected utility. Von Neumann and Morgenstern showed how to establish a ranking of the lotteries involving a set of prizes if there is: (a) a ranking of the prizes, and (b) a comparison between each prize and a lottery involving just the highest and the lowest prizes. The crucial assumption is this: Let C_1 and $C_2 n$ be the highest- and the lowest-valued prizes, respectively; then for every price C_j there is a probability $p_j (\neq 0)$ such that it is indifferent whether one chooses C_j or chooses a lottery that assigns probability p_j to C_1 and probability $1 - p_j$ to $C_2 n$. By means of substitution of indifferent lotteries and other principles, for the case of the first three lotteries in Chart 1, there would be probabilities p and q that would allow to claim the following equations, where '$\approx$' expresses sameness of value, i.e., of expected utility:

$$M_1 = ((0.5, 600), (0.5, -800))$$
$$M_2 = ((0.5, 100), (0,5, 100)) \approx ((p, 600), (1 - p, -800))$$
$$M_3 = ((1, 300)) \qquad\qquad\quad \approx ((q, 600), (1 - q, -800))$$

Clearly, as long as p and q are greater than 0.5, M_2 and M_3 are preferable to M_1. This is fine. Yet we cannot adopt this method. As long as we ought to refrain from risking bringing about the catastrophe ($C_2 n$, -800), it is immaterial whether we are considering M_2 or M_3 as alternatives. The only thing that matters is the size of the probability involved. Thus, given the overridingnesss of the catastrophe (C_2n, -800), the probabilities p and q should be both negligible and identical or at any rate threshold-identical. But it $p = q$, $M_2 \approx M_3$, which is obviously false.

It is a moot question whether there is a small probability that destroys the overridingness of a catastrophe. In purely mathematical terms one can consider a probability of the magnitude $0.00001e(1000e(1000e(1000)))$, where '$e$' means that number mentioned on the right is an exponent of the one mentioned on the left. Since a probability of this order is undoubtedly beyond our perception of probabilities, we may assume that in this discussion probability 0 and probability 1 are threshold 0 and threshold 1. That is, for purposes of choosing what one ought to perform, the overridingness of a catastrophe is destroyed only by probability (threshold) 0.

We hasten to note that the fact that classical expected utility or the Von Neumann-Morgenstern expected utility is not the value that determines what one ought morally to choose is not an objection to utility theory, or to the important applications of the theory in economics and other social sciences, or to its use in the psychological study of decision making. In these disciplines the most that is assumed is that each person acts in accordance with a utility function of his own, so that at most the theory of decision, not only utility theory, provides a *conditional* normative theory, i.e., a theory that says what the agent should choose given that he has certain values and certain utility function. Here we are concerned with an absolute point of view, and have nothing to say about the factual studies on decision making or the mathematical theories used and developed in the process of those studies.

8. *Utilitarianism, Minimax and Maximum*

Our problem of finding the best course of action is essentially a special case of the problem of finding the best strategy in a game. It is a game in which the other player is nature, i.e., the one that will choose the boa which is to be realized. However, our problem is simpler. In a two-person game there is given a matrix such that at the intersection of column j and row i lies a description or name of the pair of prizes (a_{ij}, b_{ij}) to be gained by each of the players, respectively. When probability strategies are included, the problem is to find a probability vector that will maximize gain, or alternatively, will minimize loss. In our case there is just one row of prizes and we are given *all* the probabilities that are allowed. It seems, then, that all we have to do is to find the strategy that maximizes value (or the chances of greater value), or, alternatively, to find the strategy that minimizes the risk of disvalue.

According to the *minimax* principle the best strategies are the ones that include the minimum of the maximal prizes of all the available strategies. This principle and its dual, the *maximin principle*, have the advantage that they do not require a complete ranking of the strategies. They go directly to the heart of the matter and determine the set of best strategies, without concern for the ranking of the non-best ones.

We must note that the criteria for best strategies proposed by the practitioners of decision theory apply to the values of the prizes. When the problem is to find a probability vector, this vector is to be used to compute the expected utility of the strategy.[7] But our case is different. We have seen that expected utility is not satisfactory, and, besides, we must consider values and probabilities. Nevertheless, we must turn every stone within sight in order to find a viable act-utilitarianism. We shall, therefore, apply the existing proposed criteria to the risks of value represented by the products of values and their corresponding probabilities.

Let Utilius have as set of opoas with respect to some viewpoint the set ω described in Chart 1. Then, we find the following value risks:

CHART 2
Value Risks

	C_1...	C_h...	C_i...	C_j...	C_k...	C_{2n}	maxima	minima
M_1	+300	0	0	0	−200	0	+300	−200
M_2	0	0	+50	+50	0	0	+ 50	0
M_3	0	+300	0	0	0	0	+300	0
M_4	0	+120	+30	+30	0	0	+120	0

The minimax principle yields an answer from the penultimate column. As we can see, the minimum maximum is 50; thus, according to the minimax principle alternative M_2 ought to be chosen. This is, however, incorrect, since the certainty of 300 makes M_3 the one that really ought to be chosen. In this example the maximin principle, which applies to the last column, is more obviously inadequate: it only excludes M_1 and assigns the same value risk to M_2 and M_3.

Let us restrict ourselves to the value risks of the boas which are assigned a *nonzero probability*. Then the relevant maxima and minima are:

CHART 3

	Maxima	Minima
M_1	+300	−200
M_2	+ 50	+ 50
M_3	+300	+300
M_4	+120	+ 30

Here again the minimax principle fails to make alternative M_3 the one with the highest value and, hence, the one Utilius ought to prefer. On the other hand, the maximum minimum is 300, and the maximin principle makes, correctly, action M_3 the one that Utilius ought to choose.

Yet the maximin principle is not always satisfactory. If Utilius were forced to choose just between M_2 and M_4, the maximin principle will yield the wrong answer. It will make M_2 the better action, even though M_4 with a 40% chance of producing 300 units of value is far superior.

The economist L. J. Savage has proposed an embellishment of the minimax principle. He applies this principle, not to the original matrix of a game, but to the *regret matrix* derived from it as follows: every entry a_{ij} of the original matrix is replaced with the difference $a_{ij} - A_j$, where A_j is the maximum of the column j. Taking full advantage of Savage's idea we can distinguish between a column regret matrix (as just described) and a row regret matrix (obtained by replacing a_{ij} with $a_{ij} - A_i$, where A_i is the maximum value in the ith row). Thus with the minimax and the maximin principles we get four criteria. In the case of Utilius and his set of opoas described in Chart 1, we find the following matrices: (see p. 144) Patently, the only discriminating criterion is the maximin principle applied to the row regret matrix. But even this yields an incorrect result. For it makes M_2, not M_3, the opoa Utilius ought to choose.

9. *Calibrated Expected Utility*

Act-utilitarianism does not seem, then, a going concern, for there is no

CHART 4
I. Column Regret Matrix

ω	C_1	C_h	C_i	C_j	C_k	Maxima	Minima
M_1	0	−300	−50	−50	−200	0	−300
M_2	−300	−300	0	0	0	0	−300
M_3	−300	0	−50	−50	0	0	−300
M_4	−300	−180	−20	−20	0	0	−300

I. Row Regret Matrix

ω	C_1	C_h	C_i	C_j	C_k	Maxima	Minima
M_1	0	−300	−300	−300	−500	0	−500
M_2	−50	−50	0	0	−50	0	−50
M_3	−300	0	−300	−300	−300	0	−300
M_4	−120	0	−90	−90	−120	0	−120

viable criterion of best action. Indeed, the expression 'best action' can be used as a mere synonym of 'obligatory action' or 'action member of a set from which an action ought to be chosen (or performed)'. Then in order for utilitarianism not to be an empty tautology utilitarianism must be taken to be essentially the theory about the criterion of best action or obligatoriness. But, as we have seen, the formulation of such a criterion is not an easy matter.

We shall conclude this discussion with a very tentative proposal of a formal utilitarian criterion of obligatoriness. We ourselves do not feel completely sure about it, but whatever its worth here it is. For convenience we assume that the boas determined by the relevant viewpoint are ordered in decreasing value and that the first r boas have positive value, thus:

$$C_1, ..., C_r, C_{r+1}, ..., C_{2^n},$$
where the values $v_i > 0$, for $i = 1, 2, ..., r$,
and $v_i \leq 0$, for $i = r+1, r+2, ..., 2^n$.

We need the following definitions:

D1. K is the minimal amount of disvalue regarded as a catastrophe.

D2. Ω is the subset of ω that results from ω by subtracting all opoas in ω that assign a nonzero probability to a boa C_j that $v_j \leqslant K$.

D3. Ω' is the subset of ω that results from ω by subtracting all opoas in ω that assign a nonzero probability to $C_2 n$ and $v_2 n \leqslant K$.

D4. The *negatively L-calibrated expected utility* of an opoa M_i is:

$$E_L^*(M_j) = p_1^j v_1 + \cdots + p_r^j v_r + \\ + p_{r+1}^j \frac{(Lv_i)}{(L-v_i)} + \cdots + p_{2n}^j \frac{(Lv_{2n})}{(L-v_{2n})}.$$

D5. $E_{L(m)}^*(M_j) =$ The sum of the first m terms of the above polynomial.

Now we can formulate the following:

10. Act-Utilitarian Criterion

Given a set ω of opoas from which an agent X ought to choose a course of action in a situation S at time t,

(A) X ought to choose any one of the members of the set Δ_ω to be described below, and

(B) X ought to choose none of the opoas not in Δ_ω.

The set Δ_ω is determined as follows:

(1) If there is no value K as characterized in D1, then Δ_ω is the set of all opoas in ω that have as great classical expected utility as any other member of ω;

(2) If there is a catastrophic value K and $K < v_2 n$, then Δ_ω is the set of all opoas in ω that have at least as great negatively K-calibrated expected utility as any other member of ω.

(3) If there is a catastrophic value K and $K \geqslant v_2 n$, and the set Ω is the same as ω or Ω is not empty, then Δ_ω is the set of all opoas in Ω that have at least as great negatively $v_2 n$-calibrated expected utility as any other member of Ω;

(4) If there is a catastrophic value $K \geqslant v_2 n$ and Ω is empty, but Ω' is not empty, then Δ_ω is the set of all opoas in Ω' that have at least

as great negatively $v_2 n$-calibrated expected utility as any other member of Ω';

(5) If as in (4) but Ω' is empty, then Δ_ω is the set of opoas M_j in ω for which

$$E^*_{v_2 n (2^n - 1)}(M_j) \geqslant E^*_{v_2 n (2^n - 1)}(M_h),$$

for any M_h in ω.

Indiana University

NOTES

* This paper was written while the author was a Guggenheim Fellow in 1967–68.
[1] The first round of argument appears in 'A Problem for Utilitarianism', *Analysis* **28** (1968), and the second in 'Ought, Value, and Utilitarianism', *American Philosophical Quarterly* **4** (1969).
[2] Of late we have gained a good deal of understanding into the nature of action. The recent literature on action is quite large, but among the important essays developing views in consonance with the one behind this paper we should mention: (i) H. A. Prichard, 'Duty and Ignorance of Fact', A British Academy lecture of 1932, reprinted in H. A. Prichard, *Moral Obligation*, Clarendon Press, Oxford, 1949; (ii) Arthur C. Danto, 'Basic Actions', *American Philosophical Quarterly* **2** (1965); (iii) Arthur C. Danto, 'What We Can Do', *The Journal of Philosophy* **60** (1963).
[3] An excellent pioneering discussion of the concepts of *alternative action* and *consequence* appears in Lars Bergstrom, *The Alternatives and the Consequences of Actions* Aqvist & Wiksell, Stockholm, 1966. Bergstrom gives an analysis of alternatives which does not guarantee a unique set of alternatives even if a point of view has chosen. He discusses a list of merely necessary conditions, but the result is inconclusive. Reacting to Bergstrom, Lennart Aqvist has furnished an analysis of alternativeness that guarantees uniqueness after choosing a point of view. This he does in 'Improved Formulations of Act-Utilitarianisms', *Nous* **3** (1969, forthcoming). With two-sided independence of these works, I discuss two concepts of alternative action in 'Ought, Value, and Utilitarianism'. Interestingly enough, my α-alternativeness relates to Bergstrom's and my μ-alternativeness relates to Aqvist's. None of these four concepts of alternativeness involves probability. In this regard, the analysis in Subsection 4 of this essay is an improvement. An analysis of *open action* was first given in 'Ought, Value, and Utilitarianism'.
[4] Bergstrom's 'Utilitarianism and Deontic Logic' in *Analysis* **29** (1968) is a very nice reply to my 'A Problem for Utilitarianism', that both (i) concedes the contention that value-maximizing conditions are not necessary for obligations, and (ii) proposes a recursive version of act-utilitarianism. Unfortunately, this version cannot be appraised because he relies on his analysis of alternativeness in his book, and he has furnished no sufficient condition for a set of relevant alternatives that makes a bit of his proposal feasible. See the items mentioned in Note 3. See my 'On the Problem of Formulating a Coherent Act-Utilitarianism', *Analysis* **32** (1972).

⁵ I say roughly, because the principle in question is more exactly this: "If the command 'X, do A' implies the command 'X, do B', then 'X ought to do A' implies 'X ought to do B'." The reasons for involving commands or prescriptions appear e.g., in 'Actions, Imperatives, and Obligations', *Proceedings of the Aristotelian Society*, **48** (1967–68), and in 'Acts, The Logic of Obligation, and Deontic Calculi', *Philosophical Studies* **19** (1968), and *Critica* **1** (1967).
⁶ See, e.g., R. Duncan Luce and Howard Raiffa, *Games and Decisions*, John Wiley & Sons, Inc., New York, 1957, Chapter 3.
⁷ See, e.g., Luce and Raiffa's *Games and Decisions*, Chapter 13.

ROBERT BINKLEY

COMMENTS

My discussion of Castañeda's paper[1] has two parts. In the first I commend and seek to embellish his analysis of open alternative action. In the second I defend a classical act-utilitarianism against his charge that it would lead us wrongly to risk catastrophe.

I

The notion of the action machine that Castañeda introduces is a very valuable one for clarifying our thinking about actions. I believe however that it can be made even more valuable if more patient attention is devoted to the logic of the action sentences in terms of which it is characterized. In the headlong rush to a level of abstraction suitable for logistical and mathematical treatment, certain points of philosophical importance have, I fear, been set to one side; their retrieval will enable us to define the action machine still more elegantly.

Actions are ordinarily reflected in language in subject-predicate sentences, the subject referring to the agent and the predicate characterizing the action. Typically, the predicate is built upon some distinctive verb of action: 'Utilius *stopped* the bomb', 'Utilius *rang* the alarm'. In examples of this kind the action verb is fairly specific, and conveys by itself a significant part of our knowledge of what the action is. We also have less specific, 'all purpose', action verbs which by themselves hardly convey more than that it is an action that is being described, but which can combine with other elements to yield descriptions of actions of all kinds. These all purpose verbs are of two sorts. Some, e.g., 'Utilius *caused it to be that* the bomb did not go off' require supplementation by a sentence describing a state of affairs regarded as an upshot of the agent's activity. Others, e.g., 'Utilius *performed* the act of stopping the bomb', are completed by an expression naming the action done.

That there are all purpose action verbs of this second kind shows that actions can be reflected in language in another way – in nouns or noun-like expressions rather than in verbs. A variety of reasons conspire to lead us to want to represent actions in this second way, and so to shift them from the predicates to the subjects of sentences. One important reason is that we wish to speak of classes of actions, for example, the class of right actions. Another is that we wish to use the concept of identity in connection with actions, as when we say that under the circumstances, a man's raising an eyebrow is his making a bid.

For reasons of this sort, action verbs are frequently transformed into action nouns in formalized discussions of action, and this is done by Castañeda in the present paper. We see this in his frequent use of such expressions as 'X performs A in C', in which 'A', or perhaps 'A in C' must be functioning as the name of an action. This is not done uniformly, however, since this sentence is the interpretation of his formal 'Ax', in which the 'A' must be functioning as an action predicate.

This ambivalence with respect to the syntactical category of 'A' and similar terms is not a big problem, and the sympathetic reader can easily take it in stride. Nevertheless, it *is* a *little* problem, and I believe that taking the time to clear it up is worthwhile since it will lead to a somewhat neater definition of the action machine.

That there is at least a small problem is shown by Castañeda's use of the form '$A \lor B$ is open to X in C'. Here grammar requires that the phrase '$A \lor B$' be functioning as a name, a singular term. And yet it is put together with '$\lor$', the disjunction symbol, which requires that 'A', 'B' and '$A \lor B$' all be sentences. On the other hand, the logical niceties are preserved in the formal counterpart of this which is, presumably, '$o(Ax \lor Bx)$'. Here 'A' and 'B' are action predicates, and 'o' is a sentence forming operator on sentences.

It might be objected at this point that since the difficulties I find all lie in the sentences offered as interpretation of the formalism and not in the formalism itself, the problem is at worst one of awkwardness of expression. My answer is that until we have a well-formed interpretation we do not know what the formalism means, that when we try to provide this interpretation we will be led to give up doing so with the form 'X performs A in C', and that a correct interpretation will give us an improved understanding of the action machine, and the concepts Castañeda defines in terms of it.

Of course, there is no reason why we should not represent actions in noun-like expressions. We may do so in the following way. Let us suppose that the basic form in language for talking about actions is the *simple action sentence* in which a simple action predicate is combined with a term referring to the agent. The simplicity of these action predicates consists in their not being formed out of other action predicates by any of the ordinary logical operations. 'Smokes and coughs' is thus not a simple action predicate. But there may be complexities of other sorts. For one thing, a reference to the occasion of action must always be understood. For another, these predicates have an inner logic which relates 'walks swiftly' to 'walks', for example, which I shall not attempt to discuss here.

These simple action sentences go into the notation as atomic sentences. 'Utilius exploded the bomb' becomes '$E(u, b)$', etc. We may convert these sentences describing actions to noun-like expressions naming the actions by means of an operator which I shall call the *nominalizing* operator, and represent by 'N'. It is a singular term forming operator on simple action sentences. Thus, I may use '$N[E(u, b)]$' to refer to Utilius's action of exploding the bomb (on the relevant occasion).

The nominalizing operator can be applied to false as well as to true simple action sentences. Thus it can be used to form names of merely possible, or for that matter impossible, actions as well as actual actions.

Having converted a simple action sentence into a name, we may convert it back to a sentence again by means of the all purpose action verb 'performs'. Representing this by 'P', we can represent 'Utilius performs the act of stopping the bomb' by '$P(u, N[E(u, b)])$'. One of the rules governing 'P' is, presumably, that the agent of the core action is the same as the agent of the performance. That is, such an expression as '$P(x, N[E(y, b)])$' must be rejected as ill-formed since it makes no sense to say that x performed y's action.

The nominalizing operator has so far been restricted to simple action sentences. The question naturally arises whether its range may be extended to compound action sentences. A compound action sentence is a sentence that is either a molecular compound of simple action sentences involving the same agent, or else equivalent to such a sentence in the way that 'Jones smoked and coughed' is equivalent to 'Jones smoked and Jones coughed'. If it were permitted to nominalize such sentences, then we could form the names of conjunctive, disjunctive, negative, etc. actions, and we could use the names in the same way that we use names of simple actions. We could, for example, allow ourselves to use 'X performs $A \vee B$' as an informal rendering of '$P(x, N[A(x) \vee B(x)])$'. And if this were possible, then we could systematically supply interpretations of Castañeda's formalism based on the form 'X performs A'. However, I shall argue that all this would be a mistake.

My argument is that strictly speaking there are no conjunctive, disjunctive or negative actions. There are, of course, actions that are compound in other senses. There are, for example, actions that are wholes having other actions as parts. Such compound actions can be named by listing their parts; we could do this by setting down the names of the parts separated by colons. If the parts all have a common agent, then we could speak of that action as being performed by that agent, for example, '$P(x, \{N[A(x)]:N[B(x)]\})$'. But to admit compound actions in this sense is not the same as to admit conjunctive actions. To appreciate this we may turn to the other truth-functional connectives where there is no compounding relation comparable to the part-whole relation.

Take negation. We may be tempted to use '$N[-E(u, b)]$' as the name of an action which is Utilius's not exploding the bomb. But this is not a well defined action. We are dealing here with strict logical negation, and '$-E(u, b)$' is a sentence that is true just in case it is not true that Utilius explodes the bomb. But that this action of Utilius's was not done does not imply that any other action was done, let alone an action of Utilius's. In fact, '$N[-E(u, b)]$' would not even name a well defined event since there is nothing definite that must happen if Utilius's exploding of the bomb is not to happen. Events cannot be identified by what they are not any more than men. (The composer Clemens non Papa was known as being Clemens as well as not being the Pope.) And it will not do to say that '$N[-E(u, b)]$' refers to what in fact happens when Utilius does not explode the bomb since nominalization applies to false as well as true sentences and this term would have to have a clear meaning even when in fact Utilius did explode the bomb.

Take disjunction. What action would be named by '$N[A(x) \vee B(x)]$'? Presumably it is an action which is done just in case either $N[A(x)]$ is done or $N[B(x)]$ is done. But that is to say that either it is $[NA(x)]$ or it is $[NB(x)]$, and to say this is not to specify the action. No more does one identify a man by saying that he is either the King or the Pope.

I conclude that nominalization must be restricted to simple action sentences. This means that of the kinds so far considered, the only *com-*

pound actions that can be *performed* are those that are wholes having other actions as parts. Castañeda's formalism, however, covers a much wider range, and so it is clear that it cannot be satisfactorily interpreted in terms of performance.

But this is not a particularly damaging development. The notion of performance comes into the discussion, I think, only because of the need for a single grammatical form for action sentences which leads one to draw on one of the all purpose action verbs. But if one wants to take account of truth-functional complexity, one is better off selecting a verb which is completed by sentences (which may be truth-functionally complex) rather than one completed by nounlike expressions. We have such an all purpose action verb in "cause it to be that", and an interpretation based on this will escape the objections raised against the idea of performance, and will also adapt itself very neatly to the action machine.

To make use of the idea of an agent causing something to be is to presuppose that it is possible to assign certain states of affairs to certain agents as their authors in virtue of causal relationships. These causal relationships must involve some action of the agent; otherwise there would be no ground for crediting the state of affairs to him. An agent can only cause something to happen by doing something. Thus, a causing-to-be sentence must involve a reference to an agent, to a state of affairs caused and to an action by which it was caused. Our form for this could be '$C(x, p, Ax)$', where 'x' names the agent, 'p' is a sentence describing the state of affairs caused and 'Ax' is a sentence describing the action by which the agent caused the state of affairs. The means sentence 'Ax' may itself be a causing-to-be sentence, but if so it will introduce another means sentence, and it is clear that we must eventually come to means sentences that are of other forms.

It is at this point that we may appeal to the model of the action machine. For we may suppose that the ultimate means sentences are simple action sentences describing the spinning of the dials on the action machine, and we may also suppose that these are the only simple action sentences. All other actions are to be described by the causing-to-be form.

To reconstruct the action machine on these principles, we may begin with Castañeda's atomically relevant actions, or aras. Our policy requires us to replace these with *atomically relevant states*. The *auomically relevant actions* will then be the causing to be of these states. Corresponding to

Castañeda's basic output actions there will be *basic output states*, described by conjunctions of sentences describing atomically relevant states. A *basic output action* will then be the causing to be of a basic output state.

This distinction between state and action enables us to make the point that it is the basic output *states* that are the axiological atoms, at least if we view the situation from a utilitarian perspective. For it is characteristic of utilitarianism to locate primary value in the consequences, and not in the causing of the consequences.

In the same spirit, we will have *output states* described by disjunctions of descriptions of basic output states. And there will be the corresponding *output actions* of causing the output states to be. Finally, there will be *probabilitized output states* characterized by a probability assignment to each disjunct of the output state. And there will be the *probabilitized output action* of causing this assignment of probabilities to obtain.

Thus, suppose that 'p' and 'q' are sentences characterizing the atomically relevant states, and that these are all that the viewpoint in question provides. Then there will be four basic output states given by '$p.q$', '$p.-q$', '$-p.q$', and '$-p.-q$'. There will be sixteen output states (given by the various disjunctions of these, including the null disjunction). And for each output state there will be any number of probabilitized output states depending upon the various probability assignments that the action machine provides on its dials. And for each such state there will be the corresponding probabilitized output action.

We may suppose the agent capable of spinning any dial on his machine; however, some of the dials may be inoperative. The probabilitized output states associated with operative dials may be said to be *available* probabilitized output states, and the corresponding probabilitized output actions may be said to be *open*.

Thus, Utilius may spin a certain dial, an action described by a simple action sentence. By so doing he may cause the odds to be 70 to 30 against the bomb going off. And by doing that, he may cause the bomb to go off.

II

Utilitarianism in so far as its business is to give instructions on the operation of action machines, concerns itself with probabilitized output actions, that is, acts of making certain odds obtain for and against the various basic output states. The means for doing these actions (how to spin the dials) can be left to the agent, and the actual states produced can be left to God, except in so far as the machine has dials giving probabilities of 1. All that seems to be required, then, for a viable act-utilitarianism is a means of calculating the value of a probabilitized output state from the values and probabilities of its components.

Classical utility theory provides a way of doing this. One simply computes the expected utility by computing the sum of the products of value and probability for each basic output state. The one with the greatest expected utility is the one to do. Castañeda objects to this simple minded procedure, however, on the ground that it ignores the overridingness of catastrophes. Two probabilitized output actions, he claims, might have the same expected utility, and so be morally equivalent from the point of view of our simple act-utilitarianism. Yet one of them might gain its expected utility by offering a small chance for a small value, and a lesser chance for a small loss while the other offers a dangerous gamble between a great reward on the one hand and a catastrophe on the other. Morality, he claims, forbids us to risk catastrophe, and so he concludes that we cannot go by expected utilities alone.

This argument, it seems to me, is unsound, although it rests upon a good insight. It is unsound because it presupposes an assignment of values incompatible with the claim that one of the outcomes is a catastrophe. A catastrophe is something to be avoided at all costs; it must be something, therefore, with a *very* low value, that is, with a *very large* negative value. So large that even a very small probability enables it to override other possibilities that the situation might offer. Castañeda's example of a fifty-fifty chance between a plus 600 prize and a minus 400 catastrophe is thus incoherent. A catastrophe would have to be worse than that. To put the point quite generally, classical utility theory can respond to Castañeda's objection by insisting on a different, and more realistic, assignment of values.

But there is more to it than this. Caution, or the necessity of not risking catastrophe, plays a large part in all our practical thinking, including moral thinking, and Castañeda is right to insist that our philosophizing take account of it. But considerations of catastrophe do not come into our thinking as a kind of threshhold below which we may not take chances. These considerations have rather to do with the fact that the human situation presents us with a very unbalanced range of possible values. Starting from zero, we go up only a small distance in positive values to reach the best that life has to offer. But there seems to be no limit to how far we can go in the other direction. To put numbers to it, we might say that our situation confronts us with a value range from plus 100 down to, say, minus a million or more.

Evidence for this estimate can be drawn from many sources. Compare the vivid, concrete and credible accounts of Hell given by poets with their pallid, mystical and unconvincing accounts of Heaven. Again, think of the stories in which a man purchases positive values drawn from the top of the scale at the price of negative values drawn from the bottom by selling his soul to the Devil; this always turns out to have been a bad bargain. And turning from fiction to fact, try to think what there is in the repertoire of human values that is as good for a child as burning him with napalm is bad, or what rewards life might have to offer that would be worth the extremes of pain and bereavement and other forms of ruin that are suffered by some people every day. It would appear that the only thing worth purchasing at such a price is escape from a fate still worse.

Fortunately it is possible for many people most of the time to keep the probabilities of the many possible disasters fairly low. These possibilities are so bad that once they assume a probability of any size they swamp all other considerations – we have a matter of life and death. But while these catastrophes can be avoided most of the time, the threat of them is always present, and this accounts for the emphasis that morality places on caution. This is entirely consistent with an act-utilitarianism based on classical utility theory; all that is required is that we have a just sense of the limits and dangers of life.

University of Western Ontario

NOTE

[1] 'Open Action, Utility, and Utilitarianism', this volume, p. 128.

J. S. MINAS

EMERGENT UTILITIES

I. INTRODUCTION

The point of these remarks is a terribly simple one and may be illustrated by the following crude example: An eight year old boy who prefers sexual intercourse to chocolate must surely be regarded as precocious while an eighteen year old with the reverse preferences is perverse. This example is, of course, susceptible of several kinds of objections, but the point it tries to illustrate seems an obvious one – that preferences *per se*, and their induced utilities are not the whole of the story concerning decision-making in any normative sense.

The balance of these remarks is merely a following out of this point together with some suggestions for what one hopes to be a more complete formulation of utility theory as it relates to normative decision-making.

II. NORMATIVE VS. DESCRIPTIVE

Utility theory, and its involvement in theories of decision-making, is seen here from a normative point-of-view; that is, whether or not 'good' decisions are made in the end is regarded as relevant. It is of course possible to take another point-of-view, one from which the goodness of decisions is not relevant. For example, consider an individual confronted with a fixed set of alternatives and a fixed amount of information concerning these. The problem is then set: is it possible to construct a utility function in terms of which the actual decisions of the individual may be 'explained' or rendered explicable in some way? The answer is usually 'Yes', although the explanatory model becomes increasingly complex, especially as one requires anticipatory explanations (i.e., predictions) of an individual's decision-making behaviour.

The step from explaining how one did in fact make decisions in a particular defined situation to how one would make decisions in a hypothetical particular defined situation is a natural one and according to certain theories of explanation a necessary one, perhaps even an indistinguishable one. The next step, however, to an explanation of how one *should* make decisions in an actual or hypothetical defined situation is a rather difficult one to take, and many current accounts of it seem to be confused in rather important ways.

Should you wish to concern yourself solely with descriptive accounts of what goes on, or would go on, in certain actual, or hypothetical, situations, the problems you encounter, while enormously complex, are relatively straightforward ones. That is, there are some relatively clear guides about the design of further theoretical and experimental research. However, should you wish to become concerned with normative considerations, there are no such guidelines. In fact, it is not even clear that the same theoretical concepts apply.

III. THE 'STANDARD' MODEL

I want now to turn to a brief characterisation of a generalised model of a decision-making situation. This model has five basic components, and it is not assumed here that they are independent.

D: The entity with respect to which the decision-making situation exists. D may be a psychological individual, a group of such, an organisation, etc.; it may be the operative decision-maker or the set of those on whose behalf the decision is taken.

A: The set of alternatives, precisely one of which must be selected, or the set of potential decisions, precisely one of which must be taken, or rendered actual.

B: The set of all 'utility-relevant' considerations in the situation.

C: The body of relevant information in the situation bearing on the decision.

E: A choice-criterion that identifies a subset of A, A^*, as a function of A, B, C, and D.

In typical cases where D is a human being who has a set of choices A, it is assumed that there is defined over B a dyadic relation V that is reflexive, non-symmetric, transitive, and connected; V is intended to represent a weak preference ordering. Under certain further conditions, it is frequently assumed that a numerical function U over B may be derived from V such that for each $x, y \in B$, $U(x) \geqslant U(y)$ if and only if $V(x, y)$. In such cases, C normally contains a set of conditional probabilities such that for each $y \in B$ and $x \in A$, $p(y/x)$, the conditional probability of y on x, is defined.

If all these conditions are met, E, the choice-criterion, is usually taken to be the maximisation of expected value. According to this criterion, there is defined for each $x \in A$, a number $W(x)$ as follows:

$$W(x) = \mathrm{df} \sum_{y \in B} p(y/x) \, U(y)$$

Then, A^* is defined as follows:

'$x \in A^*$' means 'There is no $y \in A$ such that $W(y) > W(x)$'.

Then, according to this criterion D did (or would, or should) choose from A^* rather than from A, and it does not matter which element of A^* is in the end selected.

In these cases, V and its related U are usually derived from behavioural studies of D. Experiments are set up in an attempt to give operational characterisations of $V(x, y)$. Where D is a group, the 'amalgamation' problem arises from attempts to synthesize the V for D from the individual V's of its members. Problems of representation arise when there are two distinct subsets of D: D_1, the decision-makers, and D_2, those on whose behalf the decision is made. In many of these cases, of course, D_1 is a proper subset of D_2.

Of course, there are numerous other interpretations of this model. In decision-making situations in the form of *games*, B is given in the pay-off matrix (expressed in terms of utilities), C contains information concerning the pay-offs to the other players, and E frequently takes the form of a 'minimax' criterion that may or may not be justified in terms of some assumptions about the other players' strategies.

In certain cases of experimental inference, the hypotheses constitute the set A and (at least in some Bayesian theories) the results of the experiment constitute one part, the *a posteriori* part, of C and another part of C, the *a priori* part, is derived from extra-experimental sources. In this connection a good bit of attention has been devoted to the sequential transformation of *a posteriori* information in one case to the *a priori* in a related succeeding case.

Approaches to decision-making from this point-of-view have been extremely successful and have had a profound impact on contemporary thinking about the general characteristics of rational behaviour. The most complex kinds of problems seem to be ultimately capable of being handled in a relatively simple way. In particular, since the individual or individuals, D, concerned in a particular situation and the available alternatives, A, seem to be given, the ultimate resolution of a particular problem seems to depend only upon B, C, and E. Particular theories of decision-making are formed around the form of B and C, and such theories are organised around some specific E; consequently, from the point-of-view of any particular theory, the problem is solvable on the basis of getting the requisite detailed information concerning B and C.

Consequently, a great deal of effort has gone into theoretical and experimental investigations of B and C, on the assumption that in this way all, or a large number of, decision problems will be solvable. It now becomes apparent that a substantial part of the motivation to do this is based on an assumption that such approaches may be used normatively rather than only descriptively. In making the step from descriptive to normative uses of decision-theory, a variety of rationales have been used. Ward Edwards, and others, have dwelt upon the concept of D's subjective perception of C, and have argued that one can describe and predict D's behaviour from knowledge of his B and his subjective C using a conventional maximisation of expectation principle. Then, the transition to a normative theory is made by replacing D's perception of C in the model by C. Along similar lines, Savage and others have argued that especially in connection with inferential problems it does not matter a great deal what one's *a priori* information is as long as an information processing system is established that will successively 'dilute' the impact of these and being into greater prominence the *a posteriori* information derived from previous controlled experiments.

IV. UTILITY

As mentioned above, the behavioural approach to utility depends upon relating D's preferences over B to his behaviour in some standardised experimental situation. On the assumption that the experimenter knows what information is available to D and that the subject is able to respond in a consistent way, it is possible to induce on these behaviourally revealed preference a utility function over B, which function is supposed to be insertible into a decision-making model concerning D that is supposed in the end to maximise D's utility or expected utility.

There are obviously many difficulties with such a behavioural approach. For example, how does the experimenter decide that B is properly defined? It is notoriously the case that in actual experiments, subjects find more utility laden aspects of the situation that the experimenter originally considered. And in the transition from the experimental situation to the actual decision-making situation, how does the investigator determine whether the utility structure has altered? How is it to be determined whether the elements of B are basic arguments of the utility function or functions themselves derived from yet more basic arguments that do not occur in B? How does the experimenter determine whether the subject's responses over time are inconsistent or differ because of changes in the subject's preferences?

These difficult methodological questions confront the investigator but do not differ in any essential way from problems arising in connection with any experimental-theoretical-applied work. But there is at least one problem that appears to be particularly associated with investigations of utility in connection with decision-making, and that is the problem of constancy of the utility function over time. One expects the human subject to exhibit both consistency and regularity on the one hand and change and development on the other. Consequently, we require a concept of preference in terms of which changes in the preference structure of the individual may be seen in some cases to be part of his 'natural' (or 'appropriate' or 'proper') development while others may be seen as evidence of inconsistency. That is, preferences and their associated utilities must not be seen as ultimates or as brute facts of the decision-making world if they are to play a significant role in that world.

It is a simple and obvious fact that decisions taken on the basis of preferences at a moment of time do have in many cases a profound influence on the formation of future preference structures. Consequently, it is irresponsible to take decisions on the assumption that no such interaction exists.

We do acknowledge the developmental character of preferences and values, as is evidenced by the plausibility of the opening example. Psychologists and sociologists as well as historians and economists also recognise this fact. But as yet, no very substantial accommodation for this has been made in the theories of those most seriously studying the character of human preferences, although the reasons for this are quite understandable.

In the first place, to take the developmental character of human preferences into account is an extremely difficult undertaking. It will undoubtedly require a very substantial complication of the model and very likely sweep into the framework a variety of considerations one would prefer to ignore (e.g., assumptions about normal human development.) And in the second place, so much effort has been required to deal with the simpler cases of static preferences in any adequate way that systematic inroads on the larger problem have not been possible.

I am only too keenly aware both of how easy it is to call for such an enrichment of the model and of how difficult it is to provide it. In the following, I'd like to present a brief review of various hypothetical attempts to deal with the problem. It will be rather sketchy and uncompelling, and I hope that soon the field will begin to develop a more substantial way of dealing with it.

V. POSSIBILITIES

The first, and perhaps most obvious, response to the problem of changing utilities is the attempt to construct a super-decision-model that is related to conventional models much in the same way that a super-game is related to a game. The 'planning horizon' concept of economists is also analogous. Here, one tries to cover a larger span of time, one within which utilities may change from stage to stage within this span. The problems here are quite obvious. For example, how does one select the length of the span? The fixing of the horizon in such approaches is known to have a substantial impact on the nature of the internal sequence of decisions. Further, how are the stage-to-stage utility changes taken into account prospectively? Are they assumed to be given by some kind of *a priori* rule (in which case one has already abandoned the sufficiency of utilitarianism) or are estimates of them derived from some other observational base (in which case the measurement and adjustment procedures are wholly unknown)? Although the 'super' approach is conceptually attractive, it appears to provide no real leg up on our problem.

A second approach, perhaps one that non- (or anti-) utilitarians have all along suspected to be necessary anyway, is to suppose that all human preferences that are susceptible to significant developmental change are for things that are really means to further ends. Then, by some mechanism or other, ultimate ends are identified, and by reference to these the whole process of utility development is rationalised and, even perhaps, controlled. The whole trouble with this tack, however, is that should it be possible to identify these ultimate ends in any operationally relevant way (i.e., a way in which their relationship to the chain of intermediate ends is made explicit), then the entire approach to decision-making that produced the problem in the first place could (and presumably should) be dispensed with altogether, except perhaps as a kind of convenient bookkeeping device.

Finally, there are attempts to provide a developmental approach to the problem of emergent preferences, within which certain patterns of development are seen to be 'preferable' to others. This may seem to some to be a begging of the question and to some others to be merely another version of the previous approach, merely substituting 'ultimate patterns' for 'ultimate ends'. I don't know whether either of these complaints is justified, but it does seem to me to be the case that if the problem of emergent preferences is to yield to study it will probably be through some such approach as this. In the end, our problem is to be able to handle the preference process, to be able to make choices between one line of development and another, and to be able to do this within a relatively non-local framework of preferences, or meta-preferences if you wish.

In the nature of the case, it seems (to me, anyway) that the appropriate locus of such a framework is in connection with the study of human beings, as psychological and social entities. Such studies have not as yet provided much benefit for decision-theory, and whether or not they are likely to do so in the future is anybody's guess. I hope they will.

University of Waterloo

ALEX C. MICHALOS

COST-BENEFIT VERSUS EXPECTED UTILITY ACCEPTANCE RULES*

1. THE PROBLEM

One of the fundamental problems of life in general and the philosophy of science in particular is the specification of reliable criteria for the determination acceptable hypotheses (in the broad sense of ordinary sentences, laws and theories) and courses of action. While no one has been able to provide a set of criteria that could be regarded as necessary and sufficient for all sorts of hypotheses and circumstances, a number of more or less plausible rules specifying sufficient conditions of acceptability *given* certain data, purposes and attitudes have obtained fairly wide acceptance. These include such familiar principles as Gauss's least squares rule of estimation, Fisher's method of maximum likelihood, Wald's minimax loss rule, Savage's minimax regret rule, Bernoulli-Bayes's rule for the maximization of expected utility, and so on.[1] The last rule in this list is of particular importance for this paper.

A number of influential philosophers have recommended the Bernoulli-Bayes rule or some variation of it as a first approximation or step in the right direction toward a solution of the problem of providing a criterion, principle or rule for determining the acceptability of scientific hypotheses (e.g., [11, 25, 27, 37, 40]). But, so far as I know, no one has suggested that some sort of benefits-less-costs rule might be more advantageous, and it is roughly this idea that I wish to explore and ultimately vindicate. More precisely, I shall attempt to prove the *normative* claim that a cost and benefit dominance principle of acceptance ought to be preferred to any sort of Bernoulli-Bayesian principle because right now and for the foreseeable future the former performs better and cannot perform worse than the latter (in a sense of 'perform' that will be elucidated below).[2]

Although most of the paper consists of a detailed analysis and comparison of the two relevant principles, their requirements and applications, I shall begin with a brief outline of the basic elements of each in order to provide a general orientation and more or less common background for our discussion.

2. MAXIMIZATION OF EXPECTED UTILITY

Proponents of the rule enjoining the maximization of expected utility, which we shall hereafter abbreviate as MEU, imagine a decision-maker confronted with a set of (practically speaking) mutually exclusive and exhaustive possible courses of action from which one that is optimal must be adopted. The decision-maker knows that the payoff or *utility* (in some sense of this word which will be explained later) that he obtains from his choice will be partially determined by events which are (practically speaking) mutually exclusive, exhaustive and beyond his control. If he has objective probability values (i.e. relative frequencies, propensities, physical range measures, etc.) for the occurrence of these events then he is operating under conditions of risk. If he does not have such values then he is in a situation of uncertainty, but he will transform it into a situation of risk by determining appropriate subjective probability values (i.e., betting quotients, degrees of belief, etc.) for the events. Given all this data he is ready to use MEU, which, as a normative principle, prescribes the acceptance of that course of action whose sum of probability-weighted utilities is larger than that of any of its alternatives. If 'U_i' and 'p_i' represent the utility and probability values, respectively, of the ith of n payoffs obtainable by adopting some hypothesis, then

(1) $\quad \sum p_i U_i \quad (i = 1, 2, ..., n)$

or

(2) $\quad p_1 U_1 + p_2 U_2 + \cdots + p_n U_n$

represents the expected utility of accepting that hypothesis. Clearly, when there is no risk involved then formulae (1) and (2) shrink to the single utility value whose procurement is certain (has a probability value of unity) given the acceptance of that hypothesis. Following MEU then, one would simply accept that hypothesis whose expected-utility (or utility in the limiting case) was greater than that of its alternatives. If two or more hypotheses have the same expected-utility and it is higher than those of the alternatives, then each of the former should be regarded as equally acceptable.

For our purposes we may think of the decision-maker described above as a practicing theoretical or applied scientist, and we may assume that his possible courses of action are the adoption of certain hypotheses as bases for further action. Although one *can*, as Isaac Levi has shown us [39], imagine a person accepting a hypothesis not as a basis for action but as somehow suitable for admission into his total corpus of knowledge, our concern here is with the provision of hypotheses more or less directly related to action (cf. [12]). If we regard a hypothesis as acceptable then, at the very least, it merits the investment of further resources such as research facilities and activity, time, money, energy, and so on. Note, however, that our concern with the acceptability of hypotheses as bases for action does not alter any of the formal or logical aspects of the problem. Only the content of the utility function or the arguments to be included in that function would be different for Levi's decision-makers and mine.

3. COST-BENEFIT DOMINANCE

Our exposition of a variation of a benefits-less-costs rule, which we will call 'the principle of cost-benefit dominance' and abbreviate CBD, may begin with a decision-maker in roughly the same situation we described for MEU. He is confronted with a similar set of (practically speaking) mutually exclusive and exhaustive hypotheses and events. He may be able to assign some sort of probability value to the occurrence of each event and he may not. In any case, such values are not required. Similarly, utility values are not required. Instead of assigning a utility value to the payoffs he will receive as a result of this or that combination of accepted hypothesis and turn of events, he merely determines the "raw forms" of the benefits and costs attached to each combination. For example, instead of noting that a hypothesis adequately accounts for a certain phenomenon provided that certain events take place rather than some others, coheres with a well-established theory in another domain given those events and has a Reichenbachian weight of .7, *and* that all of this gives it a utility value of .8, he merely lists its benefits in their "raw form" (i.e., it coheres provided that such and such is the case, etc.). Similarly, he lists its costs and the benefits and costs for its alternatives given the various contingencies. It must be assumed, of course, that he is able to weakly order the "raw form" data *within* each attribute and the corresponding preferences. For example, he must be able to determine whether two hypotheses are equally explanatory or one explains more than the other; he must be able to rank order any three distinct levels of explanatory power transitively; and he must recognize that his preferences ought to be perfectly positively correlated with the "raw form" data, e.g., he ought to prefer a hypothesis that explains more phenomena to one that explains less (if all other things are equal). It does not have to be assumed that he is able to weakly order the "raw form" data across attributes, e.g., he never has to be able to rank order different levels of explanatory power, coherence, testability, etc. on a single scale of some sort. In other words, he is obliged to make intra-attribute comparisons but not inter-attribute comparisons.

The net result of this analysis is a battery of matrices (i.e., a different matrix for each attribute) which constitute a comparative "profile" of each hypothesis with respect to its alternatives. Schematically, the matrices in such a battery for the attributes of, say, simplicity, explanatory power, precision, coherence, testability, etc. would each look like this.

Possible Events (States of Nature)

	E_1	$E_2, ..., E_n$	
H_1	$p_1 B_1^1$	$p_2 B_2^1, ..., p_n B_n^1$	

Hypotheses

H_2	$p_1 B_1^2$	$p_2 B_2^2, ..., p_n B_n^2$
⋮	⋮	⋮
H_n	$p_1 B_1^n$	$p_2 B_2^n, ..., p_n B_n^n$

The entry for row H_1 column E_2, for example, would tell us that if the hypothesis represented by 'H_1' is accepted and the event(s) represented by 'E_2' occur, then we will obtain a benefit represented by 'B_2^1' with a probability represented by 'p_2'. Since 'p_2' and 'B_2^1' need not represent numerical values of any sort, the juxtaposition of these two signs must not be taken to mean multiplication (as in formulae (1) and (2)). The superscript on 'B_2^1' indicates the hypothesis 'H_1' and the subscript indicates the event(s) 'E_2'. The 'B' is short for 'benefits'. They might be a high degree of explanatory power, simplicity, coherence with other theories, precision, etc. While various philosophers and scientists (e.g., [1, 6–8, 19, 20, 25–28, 32, 36, 40, 46–48, 53, 58, 69, 76, and 83]) have made recommendations as to which attributes ought to be included in an optimal set, all that is assumed here is that such a set would contain, say, more than a couple and less than a couple dozen members, none of which would have to be entirely independent (in any sense) of the others. In matrices for such costs as required set-up time, computational effort, special facilities, technical assistance, money, operationalization, etc. the 'B' would be replaced by a 'C' for costs'.[3] As with formulae (1) and (2), when there is no risk involved then such matrices shrink to a single item, namely, a column indicating the various benefits (costs) that will be obtained (borne) with certainty given the acceptance of a particular hypothesis.

If for every possible contingency and for every attribute, the benefits and costs of accepting one hypothesis are preferable (i.e., ought to be preferred) to those of accepting another then the former *strongly dominates* the latter. If for some contingency and for some attribute, the benefits and costs of accepting one hypothesis are preferable to those of another *and* for all of the remaining contingencies and attributes the benefits and costs of accepting the latter are not preferable to those of the former (i.e., some are exactly alike and others are less preferable), then the former dominates the latter.[4] Hence, if one hypothesis strongly dominates another then the former also dominates the latter, but the converse is not true. According to CBD then, the hypothesis that ought to be accepted is the one which dominates all of its alternatives. If two or more hypotheses have the same benefits and costs but dominate all others then they should be regarded as equally acceptable. In general, depending on the particular benefits and costs involved, research in a given problem-area should continue until some hypothesis emerges as dominant over all of its alternatives.

4. PREFERABILITY AND SUPERIOR PERFORMANCE

I am assuming that one principle of acceptance is *preferable* to another provided that the former may perform better and cannot perform worse than the latter. Moreover, one principle *performs better* than another if and only if it is more effective and more efficient than the other. An acceptance principle is *effective* exactly insofar as it is possible in every sense of this term to isolate, identify or select acceptable hypotheses by applying it. If it were impossible in any sense to apply a principle then it could not be applied and, consequently, could not identify anything. So it would be completely ineffective. The *efficiency* of an acceptance principle may be defined and measured by the ratio of its effectiveness to the number of assumptions and the degree of sophistication of the kinds of information required for its application.[5] The *degree of sophistication* of a particular piece of information may be determined by the sorts of scales (nominal, ordinal, interval or ratio) or concepts (qualitative, comparative or quantitative) required to accurately express that information, e.g., 'this is hot' may be regarded as a less sophisticated piece of information than 'this is hotter than that' which is less sophisticated than 'this has a temperature of 90°F'. Hence, if it could be shown that CBD and MEU are equally effective or that CBD is more effective than MEU *but* that CBD requires fewer assumptions and/or less sophisticated information than MEU, then the superior efficiency of CBD would be established. That, of course, would establish its superiority of performance and, therefore, its preferability over MEU, which is my central thesis. In the next two sections I shall attempt to establish this thesis roughly as follows. In Section 5 it will be shown that MEU requires information of a more sophisticated sort and, consequently, more assumptions than CBD. Some of this information is not now nor will it be in the foreseeable future available. It follows then, that MEU is now and will be for some time to come *completely ineffective*. In Section 6 we review Harvey's defense of his theory of the motion of the heart and blood agsinst that of Galen as a paradigm case study of an undoubtedly successful defense of a scientific hypothesis *and* of an implicit application of CBD. Any adequate acceptance rule would have to disclose the superiority of Harvey's view over Galen's and, as a matter of fact, it is fairly apparent that something like CBD was behind Harvey's presentation of the evidence for his view. By means of this historical example then, the *effectiveness* of CBD is established. Thus, because CBD is more effective and must have a higher efficiency ratio than MEU, the former performs better and cannot perform worse than MEU for the present and foreseeable future. Hence, the superiority and preferability of CBD over MEU is established.

5. MEU VERSUS CBD: COMPARISON OF REQUIREMENTS

The requirements and general prospects of our two principles may be thoroughly compared in five respects.

5.1. In the first place it is apparent that MEU does but CBD does not require numerical probability values. With MEU it is not enough to be able to determine that a certain event and concomitant attribute value is probable, very probable, improbable, more probable than some other, as probable as some other, etc. Such frequently useful qualitative and comparative probabilistic judgments are worthless for MEU, because the latter can only "process" quantitative judgments; e.g., judgments of the form 'The degree of probability of obtaining a value of x for attribute A in the event that E is r'. Thus, one who uses MEU must assume that all of the notoriously difficult philosophic problems involved in obtaining initial numerical probability values (for the particular procedures employed) have been satisfactorily solved. (See, for example, [5, 31, 33, 35, 49, 50, 52, 54, 56, 69, and 75].) Moreover, granted that a given position is philosophically unobjectionable in itself, it must still be assumed that every contingency that might be relevant to the acceptance of any hypothesis can be meaningfully assigned a numerical value by that procedure, i.e., that in one way or another it is always meaningful to transform conditions of uncertainty into conditions of risk. As Ellsberg has shown [5], however, proponents of minimax, maximin and maximax decision rules cannot and would not accept this assumption. (See also [10, 33 and 64].) On the other hand, both the assumption and its denial are irrelevant to CBD, for the latter does not require probability values of any sort but it can always use them when they are available.

5.2. Just as CBD can but MEU cannot get along without numerical probability values, the former can and the latter cannot proceed without numerical utility values. There seem to be two *prima facie* possible ways to produce these values, the first of which will be shown to be abortive and the second of which is largely wishful thinking. The former will be considered in this subsection and the latter in the next.

To begin, it should be noted that the numerical utility values required are cardinal and not merely ordinal. The former are necessary because they are the only ones that can be meaningfully added or multiplied, and both of these operations must be performed on the utility values used by MEU. So, some means of obtaining at least an interval scale of numerical utility values must be found.[6]

One way to tackle the problem of securing an interval scale of utility values is to try to generate them from a simple rank ordering of attribute values. The most popular and manageable means to this end is the standard-gamble technique of von Neumann and Morgenstern [65]. Unfortunately (for defenders of MEU), while this technique can (in principle though not always in fact [51]) be used to produce an interval scale of utility, the utility involved is not the sort that is of interest to philosophers of science. The latter are concerned with "epistemic utilities". In Hempel's words,

epistemic utilities... represent "gains" and "losses" as judged by reference to the objectives of "pure" or "basic" scientific research; in contradistinction to ... *pragmatic utilities*, which would represent gains or losses in income, prestige, intellectual or moral satisfaction... [26, p. 156].

And Levi writes

when an investigator declares himself to be engaged in an effort to replace agnosticism by true belief... there is no need to ascertain his "true feelings"... [40, p. 76].

Presumably the fundamental logical distinction between a scale of "epistemic utility" and "pragmatic (including psychological) utility" is that only the former could have some normative force for a scientist *as a scientist*.[7] Moreover, it might be thought that by making certain adjustments in the von Neumann-Morgenstern technique, a scale with normative force could be constructed. Indeed, this idea seems to be behind Levi's efforts in *Gambling With Truth* (p. 50). It is easy to demonstrate, however, that the von Neumann-Morgenstern technique cannot yield the required scale. According to that technique[8], a decision-maker begins by rank ordering attribute values. To keep things simple, suppose he has a single attribute, say, explanatory power, and can distinguish in a publically observable or intersubjectively testable fashion, three ranks, low, average and high. The latter may be represented by 'L', 'M' and 'H', respectively. Clearly, he prefers

H to M to L.

He then assigns a value of 0 to the lowest rank and 1 to the highest rank.

COST-BENEFIT VS. EXPECTED UTILITY ACCEPTANCE RULES 171

To determine the numerical value of the remaining rank, he considers various choices (gambles) that might be put to him having the form

$$M \text{ versus } pL + (1-p) H,$$

where 'pL' is short for 'obtaining a low ranking hypothesis with a probability value of p'. Clearly, if $p = 0$ then the right side of the gamble would be preferred (because the choice is then between M and H, and H is preferred to M), and if $p = 1$ then the left side is preferred (because M is preferred to L). By varying the value of p appropriately, the decision-maker can (often if not always) reach a point where both sides seem equally attractive. At that point he merely computes the numerical value of the right side and that tells him the value of the left side, i.e., the value of the remaining rank.

Now, in order to put this whole procedure to work *normatively*, one would have to have some criterion, rule or principle that prescribes the appropriate probability values. For the case before us, it might have the form: Given a gamble between 'M' and '$pL + (1-p) H$' a decision-maker (with his eye on the "gains" and "losses" to "pure" science) ought to reach a state of indifference when and only when $p = r$. The only way to justify such a prescription, however, is to show that M must have a certain "epistemic utility" value, and that can only be shown *after* the interval scale of "epistemic utility" values for explanatory power has been constructed. That, of course, is much too late for anyone who thought the standard-gamble technique would be useful. It is *not* useful, because one must already possess the very information it is supposed to, but can only redundantly provide. Hence, it is impossible to use this technique to construct a scale of "epistemic utility" values which would have normative force for a scientist *as* a scientist. Therefore, the first *prima facie* possible way of producing the scale required by MEU has been shown to be abortive (cf. [72, p. 5]).

5.3. The other *prima facie* plausible way to obtain the required scale is to construct interval scales of measurement for *every* relevant attribute and then transform the attribute values into utility values. Mathematically, at least, the transformations would be relatively straightforward. As a matter of fact, both Hempel and Levi have begun their constructions of "epistemic utility" measures from approximately this point. (See [26,

p. 154; and 40, p. 71]). Both of them have worked primarily *from* a "content measure" which yields values unique up to a linear transformation *to* a measure of "epistemic utility" which yields similar values. While there is nothing objectionable about this procedure in itself, it must be emphasized that it is at best a first step on a very long journey. Even if we had an acceptable measure of content, we would still almost certainly need measures of simplicity, explanatory power, precision of predictions, coherence with theories in other domains, probability, etc. in order to apply MEU. Needles to say, such measures are not available now and, judging from the history and present status of the unresolved issues surrounding the measurement of probability, they will not be available in the foreseeable future either. Again, however, such measures (interval scales) are not necessary for CBD, which may be applied as soon as one is able to construct a simple rank ordering of attribute values.

5.4. As has been suggested, the transformation of one interval scale into another is mathematically straightforward. If 'u' and 'a' represent the "epistemic utility" and attribute values, respectively, of accepting a certain hypothesis then

$$u = f(a),$$

where 'f' represents any transformation function up to and including the linear one

$$u = ma + n,$$

where 'm' and 'n' represent constants and $m = 0$. The extremely difficult problems before proponents of MEU with respect to these transformations are not, therefore, mathematical or merely technological. They are plainly philosophical. In particular, they involve the appropriate selection of 'f', or, in the linear case, the values of the constants 'm' and 'n'. In the latter case, for example, given interval scales of measurement for every relevant attribute, in order to apply MEU one must be able to systematically choose the appropriate constants and to present a plausible justification of his choices. The burden of proof then, that such and such a value for a certain attribute is worth this or that much 'epistemic utility' falls squarely upon MEU's adherents, and it is a burden which can hardly fail to create interminable haggling. After all, what principles could one invoke to show that, say, a hypothesis with a degree of simplicity of .8 ought to be assigned an "epistemic utility" of .8 or .6 or anything else? What principles could be used to prove that, say, a coherence value of .8 ought to be worth more or less than a simplicity value of .8? What could we use as a basis of comparing the significance of various attributes in order to transform their values into "epistemic utility" values? Clearly some sort of interattribute comparisons will be required to justify the transformation functions, but there is no basis for such comparisons. Of course, if we already knew the "epistemic utility" values corresponding to every attribute value for all attributes, then such comparisons would be self-evident. But that is beside the point. It is precisely those utility values that we are unable to obtain, because the required principles and bases for comparison do not exist. Indeed, even the idea that they (not to mention their justifications) might be forthcoming in the near future seems farfetched to say the least. Thus, we have another good reason for expecting perennial ineffectiveness from MEU and for turning to CBD instead.

5.5. Supposing, for the sake of argument, that all of the aforementioned problems were satisfactorily solved, proponents of MEU would still be faced with an amalgamation problem, i.e., with a problem of combining all of the individual "epistemic utility" values into a single most representative or appropriate value. Since they were forced beyond ordinal to cardinal values from the very beginning, they need not be troubled by Arrow's paradox [4, 60]. Nevertheless, some rule of combination must be constructed and its plausibility defended, and this just creates more unnecessary work in view of the availability of CBD. Because a number of amalgamation rules have already been developed, however, this final step may be the least troublesome of all. (See, e.g., [22, 29, 73, 74].) But it is still a piece of excess baggage.

To summarize the arguments in this section, I have been attempting to establish the ineffectiveness of MEU on the ground that it requires a more sophisticated sort of information than is now or will be in the foreseeable future available. To obtain this information, a number of problematic assumptions have yet to be made and substantiated. CBD, on the other hand, does not require such sophisticated information or, consequently, its attendant assumptions. Thus, a *prima facie* case for

CBD over MEU has been established. But this is not enough. It is one thing to show that CBD does not have the infelicities of MEU and another to show that CBD is effective. It is to the latter task that we shall now turn. If it can be shown that CBD is at all effective now or can be expected to be effective in the foreseeable future, then both its superior effectiveness and efficiency over MEU will be established.

6. HARVEY'S IMPLICIT USE OF CBD

In Chapter 14 of his classic *Anatomical Disquisition on the Motion of the Heart and Blood in Animals* [23], Harvey summarizes his position as follows.

Since *all things*, both argument and ocular demonstration, show that the blood passes through the lungs, and heart by the force of the ventricles, and is sent for distribution to all parts of the body, where it makes its way into the veins and porosities of the flesh, and then flows by the veins from the circumference on every side to the centre, from the lesser to the greater veins, and is by them finally discharged into the vena cava and right auricle of the heart, and this in such a quantity or in such a flux and reflux thither by the arteries, hither by the veins, *as cannot possibly be supplied by the ingesta* and *is much greater than can be required for mere purposes of nutrition*; it is absolutely necessary to conclude that the blood in the animal body is impelled in a circle, and is in a state of ceaseless motion; that this is the act or function which the heart performs by means of its pulse; and that it is the sole and only end of the motion and contraction of the heart (p. 93, italics added).

This famous paragraph is instructive in a number of ways, four of which are relevant to our discussion.

First, the fact that Harvey insists that "all things" show that his position is sound reveals an implicit acceptance of the methodological rule we have been adocating, namely, CBD. He obviously assumes that the sort of complete dominance that his hypothesis (theory) has over its rivals is a sufficient reason for accepting it. Some of the details of that dominance will be described below.

Second, the long conjunction from "show" to the italiziced phrases outlines the circular path taken by the blood through a body. This is illustrated in Figure 1 beside its most widely held alternative which was developed by Galen.[9] We shall have more to say about these diagrams below, but for now it is enough to notice that beginning with the vena cava, the arrows indicating the direction of flow in Harvey's view form a circle while those in Galen's view proceed along two straight paths

Fig. 1.

with some "back-up". This illustrates the fundamental discrepancy between the view that Harvey is advocating and the alternative that he is rejecting.

Third, the italicized phrases refer to Galen's theory that the blood is produced from ingesta and dispersed outward continuously from the liver to the rest of the body for nutrition. Although Cesalpino [13, II, pp. 226–227] had somewhat vaguely and inconsistently advocated the theory that there was a daily ebbing and flowing (like the tides) between the heart and the veins and arteries, and although there was some two-way movement in Galen's view certainly in the pulmonary and portal veins and possibly throughout the system as a result of his theory of "attractive" and "expulsive faculties" (explained below), Harvey's primary target is the idea that the blood flows more or less continuously from the liver. It is this hypothesis especially that his continuous circulation theory is to replace.

Fourth, the remarks following the italicized phrases emphasize the main aspects of his theory, namely, that the blood in animals (not just men) is moved continuously in a circle by the contraction of the heart.

A careful examination of Harvey's text suggests that Harvey considered the following attributes as especially relevant to the comparison of his theory with Galen's: explanatory power, analogies with accepted theories in other domains, simplicity and logical (internal) consistency. A priori this set of attributes has no more to recommend it than a number of others one might suggest. But for our purposes it is not necessary to reach an agreement on the optimal set of appraising attributes. All that is required here is a generally satisfactory set, or as Herbert Simon would say, a 'satisficing' set [78]. Our primary goal is to show that with respect to each of these attributes, Harvey's theory is superior to Galen's.

A. *Explanatory power*

Although Galen's theory could not account for any observable phenomena for which Harvey's theory had no explanation, the latter could but the former could not account for the facts that: (1). The amount of blood that passed into the aorta in an hour weighed much more than the total amount found in an animal [23, pp. 74–78, 87]. (2) This amount of blood did not drain all the veins or rupture the arteries [23, p. 94]. (3) It could not be produced from the juices of ingested aliment or absorbed as nutriment [23, pp. 75–78, 80, 87]. (4) When the venal cava is closed the heart becomes pale and smaller, and when the aorta is closed the heart becomes deep purple and larger [23, pp. 46–47, 79]. (5) A middling ligature closing a vein can cause a limb to swell and a tight ligature closing an artery can cause it to turn pale [23, pp. 81, 83]. (6) Dissected bodies have much more blood in their veins than in their arteries and much more in the right ventricle than the left [23, pp. 76, 111]. (7) The valves are situated in order to prevent the passage of blood from the large to the small veins which would cause swelling and rupture [23, pp. 89–90]. (8) Tumefaction follows a blow to the temple [23, p. 85]. (9) In phlebotomy a ligature must be applied above the puncture [23, p. 85]. (10) When a patient undergoing a phlebotomy becomes weaker the blood drains more slowly [23, p. 87]. (11) A whole system may become contaminated although the originally infected part is apparently sound [23, p. 96]. (12) Medicine applied externally influences internal organs [23, p. 97].

As if all of this were not enough, Harvey includes the following general remark near the end of his penultimate chapter.

> Finally, reflecting on every part of medicine, physiology, pathology, semeiotics and therapeutics, when I see how many questions can be answered, how many doubts resolved, how much obscurity illustrated by the truth we have declared,... I see a field of such vast extent in which I might proceed so far, and expatiate so widely, that this my tractate would not only swell out into a volume,... but my whole life, perchance, would not suffice for its completion [23, pp. 99–100].

B. *External analogies*

While the sort of movement envisaged by Galen's theory suggested little more than a perpetually flowing stream, the cyclic motion of Harvey's theory was analogous to and fit together admirably with a number of ideas and theories that he accepted in other domains. Indeed, immediately following his description of the path followed by the blood, he cites four analogies that would have been familiar to most of his readers.

> This motion we may be allowed to call circular, in the same way as Aristotle says that the air and the rain emulate [a] the circular motion of the superior bodies; [b] for the moist earth, warmed by the sun, evaporates; the vapours drawn upwards are condensed, and descending in the form of rain, moisten the earth again. [c] By this arrangement are generations of living things produced; [d] and in like manner are tempests and meteors engendered by the circular motion, and by the approach and recession of the sun [23, p. 71].[10]

Seven chapters later we find him again arguing in peripatetic fashion that since "under all circumstances" motion generates and preserves "heat and spirits" which are necessary for life, a body must have its "particular seat and fountain, a kind of home and hearth, where... the original of the native fire, is stored and preserved". This "original" in animals was none other than their pulsating hearts [23, p. 94; 68, p. 181].

C. *Simplicity*

Harvey's theory was simpler than Galen's in the sense that the former required fewer basic assumptions and ad hoc hypotheses than the latter. In the first place, Galen regarded the liver as the center of the venous system and the heart as the center of the arterial system, with anastomosis between the two systems and "communication between the cavities of the heart" through "the tiny pores which appear above all toward the middle of the partition between the cavities..." [23, pp. 34, 38–39; 18, p. 321]. Harvey's theory of course had a single center, the heart, and as for "the tiny pores" in the septum, he exclaimed "By Hercules! no such pores can be demonstrated, nor in fact do any such exist" [23, p. 42]. Furthermore, the ad hoc assumption of anything passing through the septum raised more questions than it answered.

> ... how could one of the ventricles extract anything from the other... when we see that both ventricles contract and dilate simultaneously? Why should we not rather believe that the right took spirits from the left, than that the left obtained blood from the right ventricle...? But it is certainly mysterious and incongruous that blood should be supposed to be most commodiously drawn through a set of obscure and invisible ducts, and air through perfectly open passages, at one and the same moment. And why... is recourse had to secret and invisible porosities, to uncertain and obscure channels, to explain the passage of the blood to the left ventricle, when there is so open a way through the pulmonary veins [23, p. 42].

It is perhaps worthwhile to notice here that although the capillaries required by Harvey's theory to permit blood to pass from arteries to veins were as "obscure and invisible" as the "uncertain and obscure channels" through the septum required by Galen's theory[11], the assumption of the existence of the former did not create more problems than it solved. Indeed, it seems to be primarily this aspect of the hypothesis of invisible capillaries which makes it decidedly not ad hoc.

Second, Galen's view of the *causes* of the movement of "material" to and from the heart was enormously more complicated than Harvey's.

The latter's view was mentioned in the fourth point that was cited above following his summary of Chapter 14. In Galen's view "almost all parts of the animal" possessed an "attractive faculty" by means of which they obtained their "proper juice", a "retentive faculty", which was responsible for the retention of whatever was of "some benefit", an "alterative faculty", which accounted for the conversion of attracted material into "nourishment", and an "expulsive faculty" that explained the elimination of whatever was not of "some benefit" [18, pp. 223–225, 247–249, 307]. For example, with respect to the stomach he explains:

...the attractive faculty in connection with swallowing, the retentive with digestion, the expulsive with vomiting and with the descent of digested food into the small intestine – and digestion itself we have shown to be a process of alteration [18, p. 275].

He also distinguishes "*two kinds of attraction*, that by which a vacuum becomes refilled and that caused by appropriateness of quality" [18, p. 319]. All "hollow organs" such as the heart and arteries display both kinds of attraction during diastole, with the former "always attracting lighter matter first" and perhaps from some distance, while the latter "acts frequently... on what is heavier" and usually nearby [18, pp. 317–319, 325].

The arteries draw into themselves on every side; those arteries which reach the skin draw in the outer air...those which pass up from the heart into the neck, and that which lies along the spine... draw mostly from the heart itself; and those which are further from the heart and skin necessarily draw the lightest part of the blood out of the veins [18, p. 317].

This should be enough for our purposes. Rather than providing a general causal explanation of the movement of "material" to and from the heart as described in his *On the Functions of Parts of the Human Body* and outlined in Figure 1, Galen has given us reasons to expect *not* that but another kind of movement. For with this explanatory scheme "almost all" of the single arrows in Figure 1 should be replaced by double arrows[12]. Furthermore, considering the fact that he only uses the term 'faculty' "so long as we are ignorant of the true essence of the cause which is operating" [18, p. 17], even if his view were internally consistent, it would not be very informative. And finally, to return to our original point, even if the scheme worked, it was much more complicated and contained many more loose ends than Harvey's.

D. *Internal consistency*

Galen was fully aware that the uncoordinated activity of the "attractive" and "expulsive faculties" could lead to paralysis or chaos. But he thought that the whole system could run smoothly if the faculties operated "consecutively" like inhaling and exhaling [18, pp. 303–307]. Harvey knew that that explanation was inadequate. If, as noted above, the "ventricles contract and dilate simultaneously" and if they only attract with dilation and repel with contraction, then they could not be exchanging anything "consecutively" [23, p. 42]. So either there was no exchange or the "faculty" scheme was faulty, or, as Harvey claimed, both. Simalarly, Harvey claimed that on Galen's view the "spirits"[13] in the aorta (which were necessary to the life of the heart as well as every other organ) should have been drawn into the left ventricle as a result of its "attractive faculty", but somehow they always escaped [23, p. 40]. This is merely a special case of the general point made above, namely, that "almost all" of the arrows on Galen's view in Figure 1 turn out to be both double and single at the same time, which is impossible. Again, Harvey saw that the idea that the mitral valve[14] should allow the "spirituous blood" to pass from the left ventricle to the lungs while at the same time it prevented the "thinner" air from retrogressing through the same channel was plainly inconsistent [23, p. 40]. And finally, he noted a similar infelicity in the alleged "cooling and cleaning system" operating between the left ventricle and lungs by means of the pulmonary vein. If the mitral valve prevented the "cooling" air from escaping once it arrived in the left ventricle then it could not fail to prevent the "fuliginous vapours" from escaping also, in which case there would be no "cleansing" activity [23, p. 40].

This completes my review of Harvey's comparison of his theory with that of Galen's on the movement of the blood and the function of the heart with respect to the four attributes of explanatory power, analogies with other theories, simplicity and internal consistency. The implicit application of CBD with these attributes was *effective*. It produced a decision in favor of Harvey's view over Galen's. Harvey was certainly not "all right", especially in his selection of acceptable scientific theories in other domains, which is quite understandable.[15] But his errors are only relevant to the content of his argument, not to its logical form. From the logical or methodological point of view, his argument was perfectly non-demonstratively valid. His theory dominated its alternatives and was, therefore' more acceptable. Fortunately (for Harvey at least), unlike most theories, his theory has continued to dominate its alternatives. In fact, by the time he and his contemporaries had passed away, serious alternatives were no longer put forward [21, pp. 172–176]. Hence, today it is appropriate to describe his theory not merely as more acceptable than its alternatives, but as acceptable.

Considering the results of Sections 5 and 6 together now, I take it that the effectiveness and efficiency, and therefore, the superiority and preferability of CBD over MEU has been established. Before closing this investigation, however, two more general topics merit our attention. The first pertains to a certain infelicity shared by both MEU and CBD, and will be discussed in the next section. The second concerns various strategies that could be used to increase the effectiveness of CBD and, therefore, strengthen our case for it. These issues are taken up in Section 8.

7. AN INFELICITY OF MEU AND CBD

The major drawback of both of these principles is that they do not provide any built-in evaluation for the variety of evidence for or against a hypothesis. If, for example, one hypothesis has a utility of .2 on the basis of a single attribute and a probability of .6 of obtaining its full value, while another hypotheses has a utility value of .2 for each of three attributes (two plus the one on which the other hypotheses is superior) with probabilities of .2 each, then the expected utility of each hypotheses is the same. 12. Similarly, neither hypothesis dominates the other. However, the hypotheses with a greater variety of support might plausibly be regarded as warranting a higher assessment. What can be said about this discrepancy?

It seems to me that this problem of assessing variety may be treated in much the same way that voting theorists treat the problem of "no election". That is, we may introduce the attribute of 'variety' into our analysis just as voting theorists introduce the option 'no election' along with the list of candidates. (See, e.g., [14].) Then, just as a voter is allowed to judge the merits of each candidate in the presence of the option to have the whole election rescinded, our decision-makers are allowed to judge the variety of support for or against a hypothesis and assign it some appropriate value. Whether or not this strategy would work as well for our decision-makers as its analogue works for voters, a priori it certainly seems that it would.

8. INCREASING THE EFFECTIVENESS OF CBD

There are a number of ways to increase the effectiveness of CBD, some of which put more severe demands on the number of assumptions and kinds of information required than proponents of CBD would be willing to satisfy. In the remaining paragraphs of this section I shall introduce five general tactics and indicate their peculiar costs.

8.1. You recall that one hypothesis was said to dominate another if *for all* attributes and contingencies the benefits and costs associated with the latter are not preferable to those of the former and for some attribute and contingency the benefits and costs associated with the former are preferable to those of the latter. The italicized phrase 'for all' may be regarded as an abbreviation of the longer locution 'for all $n(n \geqslant 1)$ relevant attributes and $m(m \geqslant 1)$ contingencies'. As long as n is sufficiently large, we may weaken the motion of dominance by degrees, by replacing 'for all n' by 'for all $n-1$', 'for all $n-2$', and so on up to 'for all $n-(n-1)$'. Thus while strong dominance and dominance place certain requirements on all attributes, $(n-1)$-dominance puts requirements on all but one attribute, $(n-2)$-dominance on all but two, etc. For example, a hypothesis whose benefits and costs were preferable to those of another with a single exception in which the latter's benefits (or costs) were preferable to the former might be said to $(n-1)$-dominate the latter, although it could not dominate the latter. Clearly, the chances of obtaining a single acceptable hypothesis with CBD increase as the degrees of dominance decrease from n. Moreover, no new information is required. On the other hand, it must be assumed that the data from some *prima facie* relevant attribute(s) may be safely ignored. That, of course, may be difficult to justify, especially in the completely general fashion proposed. For notice that, say, $(n-1)$-dominance does not specify any particular attribute to be ignored. It merely permits a reversal in any attribute whatsoever, and one may be reluctant to grant such sweeping permission.[16]

Just as one can reduce n to $n-1$, etc., one can reduce m to $m-1$, etc. when there are a sufficient number of contingencies. This would produce the same advantages and disadvantages as reductions in n.

Finally, it should be mentioned that one could again use the analogy between this investigation and voting theory, and consider such familiar notions of dominance as simple majority dominance, absolute majority dominance, $\frac{2}{3}$ dominance, and so on.[17] For specific n's and m's, all of these phrases could be translated into the 'n minus something' terminology.

8.2. Having considered the apriori elimination of attributes and contingencies, there are two fairly natural tacks to take. One may consider the elimination of specially selected contingencies or of specially selected attributes. I shall discuss the former in this subsection and the latter in the next.

It is a familiar fact that the decision rules known as 'minimax loss', 'minimax regret', 'maximin gain', 'maximax gain' and 'Hurwicz's rule' focus a dicision-maker's attention on only some of the contingencies before him. For example, minimax loss tells one to merely review the maximum losses (costs) possible as a result of accepting any hypothesis given each contingency, and to act so as to guarantee the smallest of the maximum losses possible. Similarly, one might eliminate all of the data on benefits from one's analysis and define a concept of minimax-dominance. To determine which hypothesis minimax-dominated which, one would review the maximum costs attached to each hypothesis for every contingency and regard that one as minimax-dominant which insured the smallest maximum possible cost. Concepts of minimax regret-dominance, maximax-dominance, etc. could be constructed analogously.

All of these qualified types of dominance would be easier to obtain than unqualified dominance. So their use would increase the effectiveness of CBD. Furthermore, they do not require any more information. Indeed, some of them require less information, because they completely disregard either benefits or costs. However, this demands the rather bold assumption that such data *and more* can be safely ignored, and a priori there seems to be no justification for this assumption.

8.3. As Miller and others have shown [3, 24, 62, 67, 77], there is a general tendency for decision-makers to unwittingly let one or two of many relevant attributes determine their final judgment. Moreover, as Mac-Crimmon and Raiffa [43, 72] have recently emphasized, for one reason or another, a decision-maker may *choose* to regard one attribute as more important than all of the others together. In the latter case then, concepts of specific attribute dominance might be defined such as 'probability-dominance', 'explanatory power-dominance', etc. to be used with CBD. While such an approach presupposes interattribute comparisons of importance and, therefore, additional evaluative criteria, "weighing" devices, assumptions and justifications, it is still less demanding than MEU. Hence, with this modification, CBD would probably (depending primarily on the number of vitally important attributes selected) still be more effective than MEU.

8.4. If the total expulsion of some attributes and/or contingencies from the set of relevant evidence seems unjustifiable, a less drastic procedure may seem attractive. It has already been noted that evaluative criteria and "weighing" devices would have to be developed in order to select the one or two supremely important attributes mentioned in the previous paragraph. *If* every attribute could be assigned a numerical value indicating its weight of importance *and* all attribute values could also be expressed numerically, then each hypothesis could be assigned a numerical value equal to the sum of its weighted attribute values and the hypothesis with the largest sum could be regarded as the most acceptable. By requiring all weights of importance to be real numbers greater than zero, one could be certain that every relevant attribute had *some* influence on the total evaluation sum for each hypothesis. Neat as it sounds, such a procedure could not be practicable, because it demands even more information and assumptions, and could not be more effective than MEU.

8.5. As Simon [78] and Ellsberg [15] have insisted, it is sometimes easier to determine that something is unsatisfactory than it is to determine just how satisfactory something is. It is usually easier to decide which shirts, suits, socks, or ties "just won't do" than it is to decide which of a couple fairly decent ones one should buy. Following Simon, we may say that an attribute value is satisficing if and only if hypotheses with such a value could in every sense of this term be acceptable. It follows then, that any hypothesis with a non-satisficing value for some attribute in some contingency cannot be acceptable. Hence, such hypotheses may be immediately eliminated from consideration, with CBD applied to the remainder.

By providing a good reason for rejecting some hypotheses that might otherwise remain in the set of live options, a review of attribute values from a satisficing point of view could increase the a priori chance of obtaining a single acceptable hypothesis with CBD. The apparent additional information required is the minimum satisficing attribute value for every attribute. Even if such comprehensive data was not available, however, it might still be worthwhile (i.e., increase the effectiveness of CBD) to know *some* minimum satisficing values. Naturally the identification of such values presupposes assumptions and justifications for the evaluative criteria employed.

8.6. Finally, it should be noted that one could combine some of the tactics described in 8.1–8.5.[18] For examples, one could use a satisficing review of attribute values (8.5) along with a weaker concept of dominance (8.1); a contingency eliminating rule (8.2) with a weaker concept of dominance; (8.1), (8.2) and (8.5) together; and so on. What must always be remembered, of course, is that increases in the number of tactics employed create corresponding increases in the number of assumptions and justifications required. Furthermore, the very reason CBD has been recommended here in the first place is that it is supposed to reduce the latter without excessive costs.

9. CONCLUSION

Since I have summarized the argument for my central thesis at the end of Section 4, it is not necessary to repeat it here. All that remains to be said now is that what is required at this point is a strong defense of a particular set of relevant attributes or, in Bunge's words [8], "assaying criteria" for the evaluation of all hypotheses. Perhaps no single set will do for all kinds of hypotheses. What we require of acceptable laws may be different from what we require of acceptable theories. Attributes that are relevant for the determination of the acceptability of ordinary sentences (rather than laws or theories) may well be something else again. At any rate, if my case for CBD over MEU has been argued persuasively, then it is clearly the relevant attributes or "assaying criteria" that should be the focus of our attention now.

University of Guelph, Ontario

REFERENCES

[1] Ackermann, R., 'Inductive Simplicity', *Philosophy of Science* **28** (1961) 152–161.
[2] Ackoff, R. L., *Scientific Method*, New York 1962.
[3] Archer, E. J., Bourne, L. E., and Brown, F. G., 'Concept Identification as a Function of Irrelevant Information and Instructions', *Journal of Experimental Psychology* **49** (1955) 153–164.
[4] Arrow, K. J., *Social Choice and Individual Values*, New York 1951.
[5] Barker, S. F., *Induction and Hypothesis*, Ithaca 1957.
[6] Buchdahl, G., *Metaphysics and the Philosophy of Science*, Oxford 1969.
[7] Bunge, M., *Metascientific Queries*, Springfield 1959.
[8] Bunge, M., 'The Weight of Simplicity in the Construction and Assaying of Scientific Theories', *Philosophy of Science* **28** (1961) 120–149.
[9] Bunge, M., *Scientific Research*, Vol. II, Berlin 1967.
[10] Burks, A. W., 'The Pragmatic-Humean Theory of Probability and Lewis' Theory', in *The Philosophy of C. I. Lewis* (ed. by P. A. Schilpp), LaSalle 1968, pp. 415–464.
[11] Carnap, R., *Logical Foundations of Probability*, Chicago 1950.
[12] Chisholm, R. M., 'Lewis' Ethics of Belief', in *The Philosophy of C. I. Lewis* (ed. by P. A. Schilpp), LaSalle 1968, pp. 223–242.
[13] Crombie, A. C., *Medieval and Early Modern Science*, Volume I and II, Garden City 1959.
[14] Dodgson, C. L. (Lewis Carroll), 'A Discussion of the Various Methods of Procedure in Conducting Elections', reprinted in D. Black, *The Theory of Committees and Elections*, Cambridge 1963, pp. 214–222.
[15] Ellsberg, D., Risk, Ambiguity, and the Savage Axioms, P-2173, The RAND Corporation, Santa Monica, 1961.
[16] Fishburn, P. C., *Decision and Value Theory*, New York 1964.
[17] Fleming, D., 'Galen on the Motions of the Blood in the Heart and Lungs', *Isis* **46** (1955) 14–21.
[18] Galen, *On the Natural Faculties* (trans. by A. J. Brock), London 1916.
[19] Good, I. J., 'Corroboration, Explanation, Evolving Probability, Simplicity and a Sharpened Razor', *British Journal for the Philosophy of Science* **19** (1968) 123–143.
[20] Goodman, N., 'Recent Developments in the Theory of Simplicity', *Philosophy and Phenomenological Research* **19** (1959) 429–446.
[21] Graubard, M., *Circulation and Respiration*, New York 1964.
[22] Harsanyi, J. C., 'Cardinal Welfare, Individualistic Ethics and Interpersonal Comparisons of Utility', *Journal of Political Economy* **63** (1955) 309–321.
[23] Harvey, W., *An Anatomical Disquisition on the Motion of the Heart and Blood in Animals*, Willis's translation revised and edited by A. Bowie, London, 1889 and reprinted in *Classics of Medicine and Surgery* (ed. by C. M. B. Camac), New York 1959.
[24] Hayes, J. R., 'Human Data Processing Limits in Decision Making', *Electronics System Division Report*, ESD-TDR-62-48, 1962.
[25] Hempel, C. G., 'Inductive Inconsistencies', *Synthese* **12** (1960) 439–469.
[26] Hempel, C. G., 'Deductive-Nomological Versus Statistical Explanation', in *Minnesota Studies in the Philosophy of Science*, Vol. III (ed. by H. Feigl and G. Maxwell), Minneapolis 1962, pp. 98–169.

[27] Hempel, C. G., 'Recent Problems of Induction', in *Mind and Cosmos* (ed. by R. G. Colodny), Pittsburgh 1966, pp. 112–134.
[28] Hertz, H., *The Principles of Mechanics*, New York 1956.
[29] Hildreth, C., 'Alternative Conditions for Social Orderings', *Econometrica* **21** (1953) 81–91.
[30] Hinrichs, H. H. and Taylor, G. M. (eds.), *Program Budgeting and Benefit-Cost Analysis*, Pacific Palisades, Calif. 1969.
[31] Hintikka, J. and Suppes, P., *Aspects of Inductive Logic*, Amsterdam 1966.
[32] Jeffreys, H., *Scientific Inference*, Cambridge 1957.
[33] Keynes, J. M., *A Treatise on Probability*, London 1957.
[34] Kuhn, A., *The Study of Society*, Homewood, Ill. 1963.
[35] Lakatos, I., 'Changes in the Problem of Inductive Logic', in *The Problem of Inductive Logic* (ed. by I. Lakatos), Amsterdam 1968.
[36] Laudan, L., 'Theories of Scientific Method from Plato to Mach', *History of Science* **6** (1968), 1–63.
[37] Leach, J., 'Explanation and Value Neutrality', *British Journal for the Philosophy of Science* **19** (1968) 93–108.
[38] Leinfellner, W., 'Generalization of Classical Decision Theory', in *Risk and Uncertainty* (ed. by L. Borch and J. Mossin), London 1968, pp. 196–210.
[39] Levi, I., 'On the Seriousness of Mistakes', *Philosophy of Science* **29** (1962) 47–65.
[40] Levi, I., *Gambling with Truth*, New York, 1967.
[41] Little, I. M. D., *A Critique of Welfare Economics*, Oxford 1950.
[42] Luce, R. D. and Raiffa, H., *Games and Decisions*, New York 1964.
[43] MacCrimmon, K. R., Decisionmaking Among Multiple-Attribute Alternatives: A Survey and Consolidated Approach, Memorandum RM-4823-ARPA, The RAND Corporation, Santa Monica, 1968.
[44] Mackenzie, W. J. M., *Free Elections*, London 1967.
[45] Manheim, M. L. and Hall, F. L., 'Abstract Representation of Goals', P-67-24, Department of Civil Engineering, M.I.T., 1968.
[46] Margenau, H., *The Nature of Physical Reality*, New York 1950.
[47] McLaughlin, A., 'Science, Reason and Value', *Theory and Decision* **1** (1970), 121.
[48] Michalos, A. C., *Probability and Degree of Confirmation:* A Study of the Disagreement Between Karl Popper and Rudolf Carnap from 1934 to 1964. Doctoral dissertation, University of Chicago, 1965.
[49] Michalos, A. C., 'Two Theorems of Degree of Confirmation', *Ratio* **7** (1965) 196–198.
[50] Michalos, A. C., 'Estimated Utility and Corroboration', *British Journal for the Philosophy of Science* **16** (1966), 327–331.
[51] Michalos, A. C., 'Postulates of Rational Preference', *Philosophy of Science* **34** (1967) 18–22.
[52] Michalos, A. C., 'Descriptive Completeness and Linguistic Variance', *Dialogue* **6** (1967) 224–228.
[53] Michalos, A. C., 'An Alleged Condition of Evidential Support', *Mind* **78** (1969) 440–441.
[54] Michalos, A. C., *Principles of Logic*, Englewood Cliffs 1969.
[55] Michalos, A. C., 'A Theory of Decision-Making Evaluation', paper read at the Annual Meeting of the Eastern Division of the American Philosophical Association, 1969.

[56] Michalos, A. C., 'Analytic and Other "Dumb" Guides of Life', *Analysis*, to be published.
[57] Michalos, A. C., 'Decision-Making in Committees', *American Philosophical Quarterly* **7** (1970) 91–106.
[58] Michalos, A. C., 'Positivism Versus the Hermeneutic-Dialectic School', *Theoria* **35** (1969) Part 3, 267–278.
[59] Michalos, A. C., 'The Costs of Decision-Making', *Public Choice*, to be published.
[60] Michalos, A. C., 'The Impossibility of an Ordinal Measure of Acceptability', *Philosophical Forum*, to be published.
[61] Michalos, A. C., 'Efficiency and Morality', paper read at the Annual Meeting of the Western Division of the American Philosophical Association, 1970.
[62] Miller, G. A., 'The Magical Number Seven, Plus or Minus Two', *Psychological Review* **63** (1956) 81–97.
[63] Miller, D. W. and Starr, M. K., *Executive Decisions and Operations Research*, Englewood Cliffs 1960.
[64] Milner, J., 'Games Against Nature', in *Decision Processes* (ed. by R. M. Thrall, C. H. Coombs, and R. L. Davis), New York 1960, pp. 49–60.
[65] von Neumann, J. and Morgenstern, O., *The Theory of Games and Economic Behavior*, Princeton 1947.
[66] Newman, P., *The Theory of Exchange*, Englewood Cliffs 1965.
[67] Osgood, C. S., Suci, G. J., and Tannenbaum, P. H., *The Measurement of Meaning*, Urbana 1957.
[68] Pagel, W., 'The Position of Harvey and van Helmont in the History of European Thought', in *Toward Modern Science*, Vol. II (ed. by R. M. Palter), New York 1961, pp. 175–191.
[69] Popper, K. R., *The Logic of Scientific Discovery*, New York 1959.
[70] Prest, A. R. and Turvey, R., 'Cost-Benefit Analysis: a Survey', *The Economic Journal* **75** (1965) 683–735.
[71] Pruzan, P. M., 'Is Cost-Benefit Analysis Consistent with the Maximization of Expected Utility?', in *Operational Research and the Social Sciences* (ed. by J. R. Lawrence), London 1966, pp. 319–336.
[72] Raiffa, H., Preferences for Multi-Attributed Alternatives, Memorandum RM-5868-DOT/RC, The RAND Corporation, Santa Monica, 1969.
[73] Rescher, N., *Introduction to Value Theory*, Englewood Cliffs 1969.
[74] Rothenberg, J., *The Measurement of Social Welfare*, Englewood Cliffs 1961.
[75] Salmon, W. C., *The Foundations of Scientific Inference*, Pittsburgh 1966.
[76] Schlesinger, G., *Method in the Physical Sciences*, London 1963.
[77] Shepard, R. N., 'On Subjectively Optimum Selection Among Multi-Attribute Alternatives', *Human Judgments and Optimality* (ed. by M. W. Shelly and G. L. Bryan), New York 1964, pp. 257–281.
[78] Simon, H. A. and March, J. G., *Organizations*, New York 1958.
[79] Singer, C., *A Short History of Anatomy and Physiology from The Greeks to Harvey*, New York 1957.
[80] Stedry, A. C. and Charnes, A., 'The Attainment of Organization Goals Through Appropriate Selection of Subunit Goals', in *Operational Research and the Social Sciences* (ed. by J. R. Lawrence). London 1966, pp. 147–164.
[81] Tullock, G. and Buchanan, J. M., *The Calculus of Consent*, Ann Arbor 1962.
[82] Wilkie, J. S., 'Harvey's Immediate Debt to Aristotle and to Galen', *History of Science* **4** (1965) 103–124.

[83] Williams, P. M., 'The Structure of Acceptance and Its Evidential Basis', *British Journal for the Philosophy of Science* **19** (1969) 325–344.
[84] Wilson, C. Z. and Alexis, M., 'Basic Frameworks for Decisions', *Journal of the Academy of Management* **5** (1962), 151–164.

NOTES

* The number of friends who have kindly given me suggestions and encouragement is almost embarrassingly large, but I would like to express my gratitude to Myles Brand, Cliff Hooker, David Hull, Scot Kleiner, Hugh Lehman, Werner Leinfellner, Andrew McLaughlin and Tom W. Settle.
[1] All but the last of these rules are reviewed in [54].
[2] [43] and [71] contain less thorough comparisons of these and similar rules, with respect to different applications. [72] contains a defense of the Bernoulli-Bayes principle for "multi-attribute" problems. [30] and [70] contain excellent surveys of recent work on cost-benefit analysis.
[3] A fairly thorough analysis of decision-making costs may be found in [59].
[4] These two definitions, of course, are merely special applications of the famous Pareto Principle that has been used widely by economists since Vilfredo Pareto's *Cours d'Économie politique*, 1877, e.g., [4, 22, 29, 42, 66, 74, 81].
[5] 'Effectiveness' and 'efficiency' are analyzed in greater detail in [61].
[6] On the problem of scales and their transformations see [2, 16, 42, and 74].
[7] A vast amount of literature has been produced by proponents and opponents of "pragmatic utility", and it is doubtful that I could contribute anything novel to the discussion here. Interested readers may find critiques of the concept in [4, 15, 34, 43, 45, 48, 50, 51, 74 and 80].
[8] A more thorough analysis of the technique may be found in [63].
[9] The path outlined here for Galen's view has been put together from excerpts from *On the Functions of Parts of the Human Body* in [13, 17, 21, and 79] and from the remarks of the historians themselves.
[10] Both Crombie [13, II, pp. 235–237] and Pagel [68, pp. 177–182] regard these analogies as highly influential on Harvey's thinking.
[11] The flow of blood through capillaries was not observed until 1661 by Marcello Malpighi [13, II].
[12] Crombie [13, I, pp. 164–165] and Fleming [17] completely missed this point, and criticized historians who had referred to a general ebbing and flowing in the whole venous system. See also Subsection D below.
[13] Galen imagined that the vital functions were produced by the activity of three kinds of "spirits", namely, the "vital spirit" of the heart, the "natural spirit" of the liver and the "animal spirit" of the brain. The first "accounted for" the "vital faculty" or "principle of animal life", the second for the "vegetative faculty" or "principle of nutrition and growth" and the third for the "psychic faculty" or "spiritual principle of life" [13, I, pp. 163–167].
[14] The mitral valves are located between the auricles and ventricles in the mitral orifaces on both sides of the heart. The one referred to here is on the left side and known as the 'bicuspid valve' because it has two flaps or doors.
[15] According to the first quotation in subsection B above, he evidently accepted a geocentric theory of the planetary system.

[16] Most of the case histories cited in [8] seem to have admitted some reversals, although a more careful analysis might reveal a different picture.

[17] From a formal or logico-mathematical point of view, voting theory and the theory of multi-attribute decision-making are virtually indistinguishable. See, for example, [44, 55, 57, 59 and 60].

[18] This is also suggested by MacCrimmon [43].

BIBLIOGRAPHY

Ackermann, J. C., 'Conflict and Decision', *Phil. Sci.* **34** (1967), 188–193.
Ackermann, Robert, 'Normative Explanation', *Phil. Phenomenol. Res.* **24** (1963–64), 522–9.
Ackermann, Robert, *Nondeductive Inference*, Dover, New York, 1966.
Ackoff, Russell Lincoln, 'On a Science of Ethics', *Phil. Phenomenol. Res.* **9** (1949), 663–72.
Ackoff, Russell Lincoln, *The Design of Social Research*, University of Chicago Press, Chicago, 1953.
Ackoff, Russell Lincoln, 'Games, Decisions and Organizations', *General Systems* **4** (1959), 145–50.
Ackoff, Russell Lincoln (ed.), *Progress in Operations Research*, John Wiley and Sons, New York, 1961.
Ackoff, Russell Lincoln and Churchman, C. W., 'An Experimental Definition of Personality', *Phil. Sci* **14** (1947), 304–32.
Ackoff, Russell Lincoln and Churchman, C. W., *Psychologistics*, University of Pennsylvania Press, Faculty Research Fund, Philadelphia (mimeographed).
Ackoff, Russell Lincoln and Churchman, C. W., *Methods of Inquiry*, Educational Publishers, St. Louis, 1950.
Ackoff, Russell Lincoln and Churchman, C. W., 'An Approximate Measure of Value', *Operations Research* **2** (1954), 172–80.
Ackoff, Russell Lincoln, Gupta, Shiv K., and Sayer Minas, J., *Scientific Method: Optimizing Applied Research Decisions*, John Wiley and Sons, New York, 1962.
Ackoff, Russell Lincoln and Sayer Minas, J., 'Individual and Collective Value Judgments', in M. W. Shelly and G. L. Bryan (eds.), 1964.
Ackoff, Russell Lincoln and Sasieni, Maurice W., *Fundamentals of Operations Research*, John Wiley and Sons, New York, 1968.
Adams, E. W., 'A Survey of Bernoullian Utilities and Applications', Behavioural Models Project, Technical Report No. 9, Columbia University, 1954.
Adams, E. W., 'Survey of Bernoullian Utility Theory', in H. Solomon (ed.) *Mathematical Thinking in the Measurement of Behaviour*, Free Press, Glencoe, Illinois, 1960.
Adams, E. W., 'The Logic of Conditionals', *Inquiry* **8** (1965), 166–97.
Adams, E. W. and Fagot, Robert, 'A Model of Riskless Choice', *Behavioural Science* **4** (1959), 1–10.
Agassi, J., 'The Novelty of Popper's Philosophy of Science', *Int. Phil. Quart.* **8** (1968), 442–63.
Albert, Ethel M. and Kluckhohn, C., *A Selected Bibliography on Values, Ethics, and Esthetics in the Behavioural Sciences and Philosophy* 1920–1958, University of Indiana Press, Bloomington, 1959.
Allport, Gordon, Vernon, P. E., and Lindzey, G., *Study of Values: Manual of Directions for the Study of Values*, Cambridge, Massachusetts, 1951.
Anderson, Alan Ross, 'The Logic of Norms', *Logique et Analyse* **1** (1958), 84–91.

Anderson, Alan Ross, 'Logic, Norms and Roles', *Ratio* **4** (1962), 36–49.
Anderson, Alan Ross, 'The Formal Analysis of Normative Systems', in N. Rescher (ed.), *The Logic of Decision and Action*, Pittsburgh 1967, 147–213.
Anscombe, F. J., 'Bayesian Statistics ',*The American Statistician* **15** (1961), 21–4.
Anscombe, F. J., 'Some Remarks on Bayesian Statistics', in M. W. Shelly and G. L. Bryan (eds.), (1964).
Anscombe, F. J. and Aumann, R. J., 'A Definition of Subjective Probability', *Annals of Mathematical Statistics* **34** (1963), 199–205.
Aqvist, L., 'Chisolm-Sosa Logics of Intrinsic Betterness and Value', *Nous* **2** (1968), 253–70.
Aqvist, L., 'Deontic Logic Based on a Logic of "Better"', *Acta Philosophica Fennica* **16** (Helsinki, 1963), pp. 285–90.
Armour, L., 'Value Data and Moral Rules', *Phil. Quart.* **12** (1962), 228–38.
Armstrong, W. E., 'The Determinateness of the Utility Function', *Economic Journal* **49** (1939), 453–67.
Armstrong, W. E., 'Uncertainty and the Utility Function', *Economic Journal* **58** (1948), 1–10.
Armstrong, W. E., 'Utility and the Theory of Welfare', *Oxford Economic Papers*, N.S. No. 3 (Oxford 1951), pp. 259–71.
Arnoff, E. L. and Churchman, C. W., *Introduction to Operations Research*, John Wiley and Sons, New York, 1957.
Arrow, K. J., 'Alternative Approaches to the Theory of Choice in Risk-Taking Situations', *Econometrica* **19** (1951), 404–37.
Arrow, K. J., 'Utilities, Attitudes, Choices: A Review Note', *Econometrica* **26** (1958), 1–23.
Arrow, K. J., *Social Choice and Individual Values*, 2nd ed., John Wiley and Sons, New York, 1963.
Arrow, K. J., 'Exposition of the Theory of Choice Under Uncertainty', *Synthese* **16** (1966), 253–69.
Arrow, K. J., Blackwell, D., and Girshick, M. A., 'Bayes and Minimax Solutions of Sequential Decision Problems', *Econometrica* **17** (1949), 213–44.
Audley, R. J., 'A Stochastic Model for Individual Choice Behaviour', *Psych. Rev.* **67** (1960), 1–15.
Ayer, A. J., 'Knowledge, Belief and Evidence', *Danish Yearbook of Philosophy* **1** (1964), 13–22.
Ayer, A. J., 'Induction and the Calculus of Probability', *Log. Anal.* **11** (1968), 95–144.

Baier, K. and Rescher, N., *Values and the Future*, New York 1968.
Bar-Hillel, Y., 'An Examination of Information Theory', *Phil. Sci.* **22** (1955), 86–105.
Bar-Hillel, Y. (ed.), *Language and Information*, Addison-Wesley, Reading, Massachusetts, 1964.
Bar-Hillel, Y. (ed.), *Proceedings of the 1964 International Congress for Logic, Methodology and Philosophy of Science*, North-Holland, Amsterdam, 1965.
Bar-Hillel, Y. and Carnap, Rudolf, 'Semantic Information', *Brit. J. Phil. Sci.* **4** (1953–54), 147–57.
Bartlett, M. S., 'Probability in Logic, Mathematics and Science', *Dialectica* **3** (1949), 104–13.
Batchelor, James H., *Operations Research; An Annotated Bibliography* **1** (1959), **2** (1962), **3** (1963), **4** (1964), St. Louis University Press, St. Louis.

Bates, J., 'A Model for the Science of Decision', *Phil. Sci.* **21** (1954), 326–39.
Baumer, William H., 'The One Systematically Ambiguous Concept of Probability' *Phil. Phenomenol. Res.* **28** (1968), 264–8.
Baumol, W. J., 'The Neumann-Morgenstern Utility Index: An Ordinalist View', *J. Pol. Econ.* **59** (1951), 61–6.
Baumrin, B. (ed.), *Philosophy of Science: The Delaware Seminar*, Vol. 1 (1961–62), John Wiley and Sons, New York, 1963.
Bayes, Thomas, 'An Essay Towards Solving a Problem in the Doctrine of Chances' *Phil. Trans.* **53** (1763), 370–418.
Baylis, C. A., 'Grading, Values and Choice', *Mind* **67** (1958), 485–501.
Beardslee, D. and Coombs, C. H., 'On Decision-Making Under Uncertainty', in R. M. Thrall, C. H. Coombs, and R. L. Davis (eds.), *Decision Processes*, John Wiley and Sons, New York, 1954.
Beck, Clive, 'Utterances Which Incorporate a Value Statement', *Amer. Phil. Quart.* **4** (1967), 291–9.
Becker, Gordon M., 'Sequential Decision Making: Wald's Model and Estimates of Parameters', *J. Exper. Psych.* **55** (1958), 628–36.
Becker, Gordon M. and McClintock, Charles G., 'Value: Behavioural Decision Theory', *Annual Review of Psychology*, (1967) 275–84.
Benjamin, A. C., *Science, Technology and Human Values*, University of Minnesota Press, 1965.
Benne, K. D., Bennis, W. G., and Chin, R. (eds.), *The Planning of Change*, Holt, Rinehart and Winston, New York, 1961.
Benne, Kenneth and Swanson, G. E. (eds.), 'Values and the Social Scientist', *The Journal of Social Issues* **6** (1950), 2–7.
Bennett, J. F., 'Some Aspects of Probability and Induction', *Brit. J. Phil. Sc.* **7** (1956–57), 220–30, 316–22.
Bergmann, Gustav, 'The Logic of Probability', *Amer. J. of Physics* **9** (1941), 263–72.
Bergmann, Gustav, 'Frequencies, Probabilities and Positivism', *Phil. Phenomenol. Res.* **6** (1945–46), 26–44.
Bernoulli, D., 'Specimen Theoriae Novae de Mensura Sortis', *Commentarii Academiae Scientiarum Imperialis Petropolitanae* **5** (1738), 175–92 (transl. L. Sommer), *Econometrica* **22** (1954), 23–36.
Bezembinder, T. G., Coombs, C. H., and Goode, F. M., 'Testing Expectation Theories of Decision-Making Without Measuring Utility or Subjective Probability', *J. Math. Psych.* **4** (1967), 72–103.
Binkley, R., 'A Theory of Practical Reason', *Phil. Rev.* **63** (1965), 423–48.
Black, Duncan, 'On the Rationale of Group Decision Making', *J. Pol. Econ.* **56** (1948), 23–4.
Blackwell, D. and Girshick, M. A., *Theory of Games and Statistical Decisions*, John Wiley and Sons, New York, 1954.
Blau, J. H., 'The Existence of Social Welfare Functions', *Econometrica* **25** (1957), 302–13.
Blom, Siri, 'Concerning a Controversy on the Meaning of "Probability"', *Theoria* **21** (1955), 65–98.
Bohnert, H. G., 'The Logical Structure of the Utility Concept', in R. M. Thrall, C. H. Coombs and R. L. Davis (eds.), *Decision Processes*, John Wiley and Sons, New York, 1954.
Borel, Emile, 'Probabilité et Certitude', *Dialectica* **3** (1949), 24–8.

Borel, Emile, *Probabilities and Life*, 1943. Transl. by M. Baudin, Dover, New York, 1962.
Boulding, K. E., 'Some Contributions of Economics to the General Theory of Value', *Phil. Sci.* **23** (1956), 1–14.
Braithwaite, Richard Bevin, 'The Nature of Believing', *Arist. Soc. Supp.* **33** (1932–33), 129–46.
Braithwaite, Richard Bevin, 'Belief and Action', *Proc. Arist. Soc. Supp.* **20** (1946a), 1–19.
Braithwaite, Richard Bevin, *Scientific Explanation*, 1946b. Cambridge University Press, Cambridge, 1968.
Braithwaite, Richard Bevin, 'Common Action Towards Different Moral Ends', *Proc. Arist. Soc.* **53** (1952–53), 29–46.
Braithwaite, Richard Bevin, *Theory of Games as a Tool for the Moral Philosopher*, Cambridge University Press, Cambridge, 1955.
Braithwaite, Richard Bevin, 'Why is it Reasonable to Base a Betting Rate Upon an Estimate of Chance?' Yehoshua Bar-Hillel (ed.), *Logic Methodology and Philosophy of Science* (1967), pp. 263–73.
Braybrooke, David, 'Let Needs Diminish that Preference May Prosper', in *Studies in Moral Philosophy* (Oxford 1968), *Amer. Phil. Quart.*, Monograph, No. 1.
Braybrooke, David and Lindblom, Charles E., *A Strategy of Decision*, New York, 1963.
Brim, Orville Gilbert, *Personality and Decision Processes; Studies in the Social Psychology of Thinking*, Stanford University Press, Stanford, 1962.
Broad, C. D., 'The Principles of Problematic Induction', *Proc. Arist. Soc.* **28** (1938).
Brody, B. D., 'Confirmation and Explanation', *J. Phil.* **65** (1968), 282–99.
Bronowski, Jacob, *Science and Human Values*, New York 1958.
Bross, I., *Design for Decision*, Macmillan, 1953.
Brown, George Spencer, *Probability and Scientific Inference*, Longmans, London, 1957.
Brownson, H. C., 'Research on Handling Scientific Information', *Science* **132** (1961), 1922–31.
Bryan, Glenn L. and Shelly, M. W., *Human Judgments and Optimality*, John Wiley and Sons, New York, 1964.
Buchanan, J. M. and Tulloch, G., *The Calculus of Consent*, Ann Arbor 1962.
Buchanan, J. M., 'Individual Choice in Voting and the Market', *J. Pol. Econ.* **62** (1954a), 334–43.
Buchanan, J. M., 'Social Choice, Democracy, and Free Markets', *J. Pol. Econ.* **62** (1954b), 114–23.
Bush, R. R., Luce, R. D., and Galanter, E., *Handbook of Mathematical Psychology III*, John Wiley and Sons, New York, 1965.
Butts, Robert E., 'Feyerabend and the Pragmatic Theory of Observation', *Phil. Sc.* **33** (1966), 383–94.

Carnap, Rudolf, 'The Two Concepts of Probability', *Phil. Phenomenol. Res.* **5** (1944–45), 513–32.
Carnap, Rudolf, 'On the Application of Inductive Logic', *Phil. Phenomenol. Res.* **8** (1947–48), 133–47.
Carnap, Rudolf, *Logical Foundations of Probability*, 1950. 2nd ed., University of Chicago Press, 1962.
Carnap, Rudolf, *The Continuum of Inductive Methods*, University of Chicago Press, 1962.

Carnap, Rudolf, *The Philosophy of Rudolf Carnap*, ed. Paul Arthur Schilpp, Cambridge University Press, 1963a.
Carnap, Rudolf, 'The Philosopher Replies: (V): Probability and Induction', in P. A. Schilpp (ed.), *The Philosophy of Rudolf Carnap*, 1963b.
Carnap, Rudolf, *Logic and Language: Studies Dedicated to Rudolf Carnap on his 70th Birthday*, D. Reidel, Dordrecht-Holland, 1962a.
Carnap, Rudolf, 'The Aim of Inductive Logic', in Nagel *et al.* (eds.), (1962b), 303–18.
Carter, C. F., Meredith, G. P., and Shackle, G. L. S., *Uncertainty and Business Decisions*, Liverpool University Press, 1957.
Cashen, Walter E., 'War Games and Operations Research', *Phil. Sci.*, 22 (1955), 309–20.
Castañeda, Hector, 'Imperatives, Decisions and "Oughts": A Logico-Metaphysical Investigation', in H. Castañeda and G. Nakhnikian (eds.), *Morality and the Language of Conduct*, Wayne State University Press, 1965, pp. 219–99.
Castañeda, Hector, 'A Problem for Utilitarianism', *Analysis*, 28 (1968).
Castañeda, Hector, 'Ought, Value and Utilitarians', *Amer. Phil. Quart.*, 6 (1969).
Caws, Peter, 'The Paradox of Induction and the Inductive Wager', *Phil. Phenomenol. Res.* 22 (1961–62), 512–20.
Caws, Peter, *Science and the Theory of Value*, New York 1967.
Cerf, Walter, 'Value Decisions', *Phil. Sci.* 18 (1951), 26–34.
Chapman, J. S., 'Foundations of Utility', *Econometrica* 28 (1960), 223–4.
Chernoff, H., 'Rational Selection of Decision Functions', *Econometrica* 22 (1954), 422–43.
Chernoff, Herman and Moses, L., *Elementary Decision Theory*, John Wiley and Sons, New York, 1959.
Cherry, Colin E., 'A History of the Theory of Information', *Methodos* 8 (1956), 57–91.
Chisholm, Roderick, 'The Descriptive Element in the Concept of Action', *J. Phil.* 61 (1964), 613–25.
Chisholm, R. and Sosa, E., 'On the Logic of Intrinsically Better', *Amer. Phil. Quart.* 3 (1966), 244–9.
Chisholm, Roderick, 'What is it to Act Upon a Proposition?' *Analysis* 22 (1961), 1–6.
Chisholm, Roderick and Sosa, E., 'Intrinsic Preferability and the Problem of Superogation', *Synthese* 16 (1966), 321–31.
Church, Alonzo, 'On Carnap's Analysis of Statements of Assertion and Belief', *Analysis* 10 (1950), 97–9.
Churchman, C. West, 'Probability Theory', *Phil. Sci.* 12 (1945), 147–73.
Churchman, C. West, *Theory of Experimental Inference*, Macmillan, New York, 1948.
Churchman, C. West, 'Reply to Comments on "Statistics, Pragmatics, Induction"', *Phil. Sci.* 16 (1949), 151–3.
Churchman, C. West, 'Problems of Value Measurement for a Theory of Induction and Dcisions', in *Proceedings of the 3rd Berkeley Symposium on Mathematical Statistics and Probability*. University of California Press, Berkeley, 1955, pp. 53–9.
Churchman, C. West, 'Science and Decision Making', *Phil. Sci.* 23 (1956), 247–9.
Churchman, C. West, *Prediction and Optimal Decision: Philsosophical Issues of a Science of Values*, Prentice-Hall, Englewood Cliffs, New Jersey, 1961a.
Churchman, C. West, 'A Pragmatic Theory of Induction', in P. G. Frank (ed.), *The Validation of Scientific Theories*, 1961b.
Churchman, C. West, 'Decision and Value Theory', in R. L. Ackoff (ed.), *Progress in Operations Research I*, John Wiley and Sons, New York, 1961.
Churchman, C. West, 'Statistics, Pragmatics, Induction', *Phil. Sci.* 15 (1948).

Churchman, C. West and Arnoff, E. L., *Introduction to Operations Research*, John Wiley and Sons, New York, 1957.
Churchman, C. West and Ratoosh, P. (eds.), *Measurement: Definitions and Theories*, John Wiley and Sons, New York, 1959.
Cohen, J. and Hansel, C. E. M., *Risk and Gambling: the Study of Subjective Probability*, Philosophical Library, 1956.
Cohen, L. Jonathan, 'A Logic for Evidential Support', *Brit. J. Phil. Sci.* **17** (1966–67), pp. 27–43, 105–26.
Cohen, R. and Wartofsky, M., *Boston Studies in the Philosophy of Science*, II, Humanities Press, New York, 1965.
Colodny, Richard, (ed.), *Mind and Cosmos*. University of Pittsburgh Press, Pittsburgh, 1966.
Comprehensive Bibliography on Operations Research, Operations Research Group, Case Institute of Technology. John Wiley and Sons, New York, 1963.
Coombs, Clyde H., 'Mathematical Models in Psychological Scaling', *Journal of the American Statistical Association* **46** (1951), 480–9.
Coombs, Clyde H., 'Psychological Scaling Without a Unit of Measurement', *Psych. Rev.* **57** (1950), 145–58.
Coombs, Clyde H., 'Social Choice and Strength of Preferences', in R. M. Thrall, C. H. Coombs, and R. L. Davis (eds.), *Decision Processes*, John Wiley and Sons, New York, 1954.
Coombs, C. H. and Beardslee, D., 'On Decision-Making Under Uncertainty', in R. M. Thrall, C. H. Coombs, and R. L. Davis (eds.), *Decision Processes*, John Wiley and Sons, New York, 1954.
Coombs, C. H., Bezembinder, T. G., and Goode, F. M., 'Testing Expectation Theories of Decision-Making Without Measuring Utility or Subjective Probability' *Journal of Mathematical Psychology* **4** (1967), 72–103.
Coombs, C. H. and Komorita, S. S., 'Measuring Utility of Money Through Decisions', *Amer. J. Psych.* **71** (1958), 383–9.
Coombs, C. H. and Pruit, D. G., *A Study of Decision Making Under Risk*, Report No. 2900-33-T, Willow Run Laboratories, University of Michigan, 1960.
Coombs, C. H., Thrall, R. M., and Davis, R. L. (eds.), *Decision Processes*, John Wiley and Sons, New York, 1954.
Cooper, Neil, 'The Concept of Probability', *Brit. J. Phil. Sci.* **16** (1966–67), 226–38.
Copeland, Arthur H., 'Statistical Induction and the Foundations of Probability'. *Theoria*, **28** (1962), 27–44, 87–109.
Cowan, Thomas A., 'A Note on Churchman's "Statistics, Progmatics, Induction"', *Phil. Sci.* **16** (1949), 148–51.
Cramer, Harald, *The Elements of Probability Theory and Some of Its Applications*, John Wiley and Sons, New York, 1955.
Cunningham, R. L., 'Inductive Ascent the Same as Inductive Descent', *Mind* **72** (1963), 598.

Danto, Arthur, 'Basic Actions', *Amer. Phil. Quart.* **2** (1965), 141–8.
Davidson, Donald, 'Actions, Reasons and Causes', *J. Phil.* **60** (1963), 685–700.
Davidson, Donald and Marschak, J., 'Experimental Tests of a Stochastic Decision Theory', in C. W. Churchman and P. Ratoosh (eds.), 1959.
Davidson, Donald, McKinsey, J. C. C., and Suppes, P., *Outlines of a Formal Theory of Value*, Vol. I., Stanford, 1954.

Davidson, Donald and Suppes, P., 'A Finitistic Axiomatization of Subjective Probability and Utility', *Econometrica* **24** (1956), 264–75.
Davidson, Donald, Suppes, P., and Seigel, S., 'Some Experiments and Related Theory in the Measurement of Utility and Subjective Probability', Report No. 4, Stanford Value Theory Project, 1955.
Davidson, Donald, Suppes, P., and Seigel, S., *Decision Making: An Experimental Approach*, Stanford, 1957.
Davis, J. M., 'The Transitivity of Preferences', *Behavioural Science* **3** (1958), 26–33.
Day, J. P., *Inductive Probability*, 1948, Humanities Press, New York, 1961.
Debreu, G., 'Representation of a Preference Ordering by a Numerical Function', in R. M. Thrall, C. H. Coombs, and R. L. Davis (eds.), *Decision Processes*, John Wiley and Sons, New York, 1954.
Debreu, G., 'Stochastic Choice and Cardinal Utility' *Econometrica* **26** (1958), 440–4.
Debreu, G., *Theory of Value*, John Wiley and Sons, New York, 1959.
De Finetti, Bruno, 'Le vrai et le probable', *Dialectica* **3** (1949), 78–92.
De Finetti, Bruno (ed.), 'Foundations of Probability', in *Philosophy in the Mid-century*, La Nuova Italia Editrice, Florence, 1958.
De Finetti, Bruno, 'Probability: Philosophy and Interpretation', mimeographed, 1963.
De Finetti, Bruno, 'La prévision; ses lois logiques, ses sources subjectives', *Ann. Hist. H. Poincaré* **7**, 1–68.
De Groot, M. H., 'Some Comments on the Experimental Measurement of Utility', *Behavioural Science* **8** (1963), 146–9.
De Morgan, Augustus, *Theory of Probabilities*, 1849.
Dewey, John, *Theory of Valuation, International Encyclopedia of Unified Science*, Vol. II, p. 4, Chicago University Press, Chicago, 1939.
Diesing, Paul, *Reason in Society: Five Types of Decision and Their Social Conditions*, University of Illinois Press, 1963.
Donaldson, William, White, D., and Lawrie, N., *Operational Research Techniques*, Business Books, London, 1969.
Dyckman, J. W., 'Planning and Decision Theory', *American Institute of Planning Journal* **27** (1961), 343–5.

Edginton, E. S., 'On the Possibility of Rational Inconsistent Beliefs', *Mind* **77** (1968), 582–3.
Edgley, Roy, 'Practical Reason', *Mind* **294** (1965), 174–91.
Edel, Abraham, 'Concept of Values in Contemporary Philosophical Value Theory', *Phil. Sci.* **20** (1953), 198–207.
Edwards, Ward, 'Probability Preferences in Gambling', *Amer. J. Psych.* **66** (1953), 349–64.
Edwards, Ward, 'Probability Preferences Among Bets with Different Expected Values', *Amer. J. Psych.* **67** (1954a), 56–67.
Edwards, Ward, 'The Reliability of Probability Preferences', *Amer. J. Psych.* **67** (1954b), 68–95.
Edwards, Ward, 'The Theory of Decision Making', *Psychological Bulletin* **51** (1954c), 380–417.
Edwards, Ward, 'Note on Potential Surptise and Nonadditive Subjective Probabilities', in M. J. Bowman (ed.), *Expectations, Uncertainty and Business Behaviour*, Social Science Research Council, 1958.
Edwards, Ward, *Subjective Probability in Decision Theory*, Report No. 2144-361-T,

Project Michigan, Willow Run Laboratories, University of Michigan, 1959.
Edwards, Ward, 'Behavioural Decision Theory', *Annual Review of Psych.* **12** (1961), 473–98.
Edwards, Ward, 'Utility, Subjective Probability, Their Interaction and Variance Preferences', *J. Conflict Resolutions* **6** (1962), 42–51.
Edwards, Ward, Lindman, H., and Savage, L. J., 'Bayesian Statistical Inference for Psychological Research', *Psych. Rev.* **70** (1963), 193–242.
Ellis, Brian, 'A Vindication of Scientific Inductive Practices', *Amer. Phil. Quart.* **2** (1965), 296–304.
Ellis, R. L., 'On The Foundations of the Theory of Probabilities', *Transactions of the Cambridge Philosophical Society*, **8** (1844), pp. 1–6.
Ellsberg, D., 'Classical and Current Notions of "Measurable Utility"', *Economic Journal* **64** (1954), 528–56.
Ellsberg, D., 'Risk, Ambiguity and the Savage Axioms', *Quarterly Journal of Economics* **75** (1961), 643–69.
Emmerich, D. S. and Greene, J., 'Some Decision Factors in Scientific Investigation', *Phil. Sci.* **33** (1966), 262–327.
Ethanbolker, 'Simultaneous Axiomatization of Utility and Subjective Probability', *Phil. Sci.* **34** (1967), 333–40.

Fagot, Robert and Adams, E. W., 'A Model of Riskless Choice', *Behavioural Science* **4** (1959), 1–10.
Feather, N. T., 'Subjective Probability and Choice Behaviour', *J. Exper. Psych.* **58** (1959), 257–66.
Feather, N. T., 'Subjective Probability and Decision Under Uncertainty', *Psych. Rev.* **66** (1959a), 195–264.
Feibleman, James, 'Pragmatism and Inverse Probability', *Phil. Phenomenol. Rev.* **5** (1944–45), 309–19.
Feigl, Herbert, 'The Difference Between Knowledge and Valuation', in K. Benne and G. E. Swanson (eds.), 1950.
Feigl, Herbert, 'Validation and Vindication', in W. Sellars and J. Hospers (eds.) 1952.
Feigl, Herbert and Brodbeck, May, *Readings in the Philosophy of Science*, Appleton-Century-Crofts, New York, 1953.
Feigl, Herbert and Scriven, Michael, *Minnesota Studies in the Philosophy of Science*, Vol. 1 (1956).
Feigl, Herbert, Scriven, Michael, and Maxwell, Grover, *Minnesota Studies in the Philosophy of Science*, Vol. 2 (1958).
Feigl, Herbert and Maxwell, Grover, *Current Issues in the Philosophy of Science*, Holt, Rinehart and Winston, New York, 1961.
Feigl, Herbert and Maxwell, Grover, *Minnesota Studies in the Philosophy of Science*, Vol. 3 (1962).
Feller, William, *An Introduction to Probability Theory and Its Applications*, Vol. II, John Wiley and Sons, New York, 1966.
Fellner, William, *Probability and Profit: A Study of Economic Behaviour Along Bayesian Lines*, Richard D. Erwin Inc., 1965.
Fellows, Erin W., 'Social and Cultural Influences in the Development of Science', *Synthese* **13** (1961), 156–72.
Feraud, L., 'Induction Amplifiante et Inférence Statistique', *Dialectica* **3** (1949), 127–52.
Feyerabend, P. K., 'A Note on the Problem of Induction', *J. Phil.* **61** (1964), 349–53.

Feyerabend, P. K., 'A Note on Two Problems of Induction', *Brit. J. Phil. Sc.* **19** (1968), 251–3.
Finch, Henry Albert, 'An Explication of Counterfactuals by Probability Theory', *Phil. Phenomenol. Res.* **18** (1957–58), 368–78.
Finch, Henry Albert, 'Confirming Power of Observations for Decisions Among Hypotheses', *Phil. Sci.* **27** (1960), 293–307.
Fine, Arthur A., 'Consistency, Derivability and Scientific Change', *J. Phil.* **64** (1967), 231–40.
Fine, Arthur A., 'Probability, Logic and Quantum Theory', *Phil. Sci.* **35** (1968), 101–11.
Fishburn, P. C., 'A Normative Theory of Decision Under Risk', Ph.D. Thesis, Case Institute of Technology, 1961.
Fishburn, P. C., *Decision and Value Theory*, John Wiley and Sons, New York, 1964.
Fishburn, P. C., 'Decision Under Uncertainty: An Introductory Exposition', *The Journal of Industrial Engineering* **17** (1966), 341–53.
Fishburn, P. C., 'Preference-Based Definitions of Subjective Probability', *Annals of Mathematical Statistics* **38** (1967a).
Fishburn, P. C., 'Methods of Estimating Additive Utilities', *Management Science* **13** (1967b), 450–3.
Fisher, Ronald Aylmer, *Statistical Methods and Scientific Inference*, 2nd ed., Oliver and Boyd, Edinburgh, 1959.
Flood, M. M., 'Game Learning Theory and Some Decision Making Experiments', in R. M. Thrall, R. M. Coombs and R. L. Davis (eds.), *Decision Processes*, John Wiley and Sons, New York, 1954.
Foster, M. H. and Martin, M. L., (ed.), *Probability, Confirmation and Simplicity*, The Odyssey Press, New York, 1966.
Frank, Phillip G. (ed.), *The Validation of Scientific Theories*, Collier Books, New York, 1961.
Frankena, W. K., 'G. H. Wright on the Theory of Morals, Legislation and Value', *Ethics* **76** (1966), 131–6.
Freudenthal, Hans, 'Models in Applied Probability', *Synthese* **12** (1960), 202–12.
Freund, John E., 'Statistical versus Pragmatic Inference', *Phil. Sci.* **16** (1949), 142–7.
Friedman, M. and Savage, L. J., 'The Utility Analysis of Choices Involving Risks', *Journal of Political Economy* **56** (1948), 279–304.
Friedman, M. and Savage, L. J., 'The Expected-Utility Hypothesis and the Measureability of Utility', *Journal of Political Economy* **60** (1952), 463–74.
Fry, T. C., 'A Mathematical Theory of Rational Inference', (a Non-Mathematical Discussion of Baye's Theorem), *Scripta Mathematica* **2** (1934), 205–21.

Galanter, E., 'The Direct Measure of Utility and Subjective Probability', *Amer. J. Psych.* **75** (1962), 208–20.
Gauthier, David P., *Practical Reasoning*, Oxford, 1964.
Gauthier, David P., 'How Decisions are Caused', *J. Phil.* **64** (1967), 147–51.
Gauthier, David P., 'How Decisions are Caused (But not Predicted)', *J. Phil.* **65** (1968), 170–1.
Gini, C., 'Concept et Mesure de la Probabilité', *Dialectica* **3** (1949), 36–54.
Girshick, M. A. and Blackwell, D., *Theory of Games and Statistical Decisions*, John Wiley and Sons, New York, 1954.
Golightly, C. I., 'Value as a Scientific Concept', *J.Phil.* **53** (1956), 233–45.
Good, Irving John, *Probability and the Weighing of Evidence*, Hafner, New York, 1950.

Good, Irving John, 'Rational Decision', *Journal of the Royal Statistical Society* (B), **14** (1952), 107–14.
Good, Irving John, 'Kinds of Probability', *Science* **129** (1959), 443–7.
Good, Irving John, 'The Paradox of Confirmation, I', *Brit. J. Phil. Sci.* **11** (1960–61), 145–9.
Good, Irving John, 'The Paradoxes of Confirmation, II', *Brit. J. Phil. Sci.* **12** (1961–62), 63–4.
Good, Irving John, *The Estimation of Probabilities*, MIT Press, Cambridge, Massachusetts, 1965.
Good, Irving John, 'On the Principle of Total Evidence', *Brit. J. Phil. Sci.* **17** (1966–67), 319–21.
Good, Irving John, 'The White Shoe is a Red Herring', *Brit. J. Phil. Sci.* **17** (1966–67b), 322.
Good, Irving John, 'Corroboration, Explanation, Evolving Probability, Simplicity and a Sharpened Razor', *Brit. J. Phil. Sci.* **19** (1968), 123–43.
Goodman, Nelson, 'Some Reflections on the Theory of Systems', *Phil. Phenomenol. Res.* **9** (1948–49), 620–5.
Goodman, Nelson, *Fact, Fiction and Forecast*, Harvard University Press, 1955. 2nd ed., Bobbs-Merrill, 1965.
Gore, W. S. and Silander, F. S., 'A Bibliographical Essay on Decision-Making', *Administrative Science Quarterly* **4** (1959), 97–121.
Gotshalk, D. W., 'Value Science', *Phil. Sci.* **19** (1952), 183–92.
Gray, P. and Machol, R. E. (eds.), *Recent Developments in Information and Decision Processes*, Macmillan, New York, 1962.

Haas, W., 'Value Judgments', *Mind* **62** (1953), 512–7.
Hacking, Ian M., 'Guessing by Frequency', *Proc. Arist. Soc.* **64** (1963–64), 55–70.
Hacking, Ian M., 'On the Foundations of Statistics', *Brit. J. Phil. Sci.* **15** (1964–65), 1–26.
Hacking, Ian M., *The Logic of Statistical Inference*, Cambridge University Press, Cambridge, 1965.
Hacking, Ian M., Review of *Gambling with Truth*, I. Levi, *Synthese* **17** (1967), 444–8.
Hacking, Ian M., 'Slightly More Realistic Personal Probability', *Phil. Sci.* **34** (1967), 305–10.
Hadley, G., *Introduction to Probability and Statistical Decision Theory*, Holden Day, San Francisco, 1967.
Hall, Everett W., *Modern Science and Human Values*, Princeton 1957.
Hall, Everett W., *Our Knowledge of Fact and Value*, Chapel Hill 1961.
Hallden, Soren, 'Preference Logic and Theory Choice', *Synthese* **16** (1966), 307–32.
Hallden, Soren, 'On the Logic of "Better"', Uppsala 1957.
Halmos, P. R., *Measure Theory*, Van Nostrand, New York, 1950.
Hamblin, C. L., 'The Modal "Probably"', *Mind* **68** (1959), 234–40.
Hammerton, M., 'Bayesian Statistics and Popper's Epistemology', *Mind* **77** (1968), 109–17.
Hampshire, Stuart and Hart, H. L. A., 'Decision, Intention and Certainty', *Mind* **67** (1958), 1–12.
Hannsson, Bengt, *Topics in the Theory of Preference Relations*, Lund, 1964.
Hanson, N. R., 'The Logic of Discovery', *J. Phil.* **55** (1958), 1013–69.
Hanson, N.R., 'Is There a Logic of Scientific Discovery?', *Aust. J. Phil.* **38** (1960), 91–106

Hanson, N. R., 'The Idea of a Logic of Discovery', *Dialogue* **4** (1965), 48–61.
Harman, Gilbert, 'The Inference to the Best Explanation', *Phil. Rev.* **74** (1965).
Harman, Gilbert, 'Detachment, Probability and Maximum Likelihood', *Noûs* **1** (1968a), 401–44.
Harman, Gilbert, 'Enumerative Induction as Inference to the Best Explanation', *J. Phil.* **65** (1968b), 529–33.
Harman, Gilbert, 'Knowledge, Inference and Explanation', *Amer. Phil. Quart.* **5** (1968), 164–73.
Hartman, Robert S., 'Value, Fact and Science', *Phil. Sci.* **25** (1958), 97–108.
Harrah, David, 'Science and the Rhetorical Aspect of Communication', *Methodos* **9** (1957), 113–22.
Harrah, David, 'A Model for Applying Information and Utility Functions', *Phil. Sci.* **30** (1963), 267–73.
Harrod, Sir Roy, 'The General Structure of Inductive Argument', *Proc. Arist. Soc.* **61** (1960–61), 41–56.
Harsanyi, John C., 'Cardinal Welfare, Individualistic Ethics, and Interpersonal Comparisons of Utility', *J. Pol. Econ.* **63** (1955), 309–21.
Hartman, R., 'Formal Axiology and the Measurement of Values', *J. Value Inq.* **1** (1967), 38–46.
Heidelberger, Herbert, 'Knowledge, Certainty and Probability', *Inquiry* **6** (1963), 245–55.
Hempel, Carl G., 'Inductive Inconsistencies', *Synthese* **12** (1960), 439–469.
Hempel, Carl G., 'Rational Action', *Proceedings and Addresses of the APA*, **35** (1961–62), 5–23.
Hempel, Carl G., *Aspects of Scientific Explanation*, Free Press, New York, 1965a.
Hempel, Carl G., 'Coherence and Morality', *J. Phil.* **62** (1965b), 539–42.
Hempel, Carl G., 'The White Shoe: No Red Herring', *Brit. J. Phil. Sci.* **18** (1967–68), 239–40.
Hempel, Carl G., 'Deductive-Nomological versus Statistical Explanations', in H. Feigl and G. Maxwell (eds.), *Minnesota Studies in the Philosophy of Science*, Vol. 3.
Hintikka, Jaakko, 'Induction by Enumeration and Induction by Elimination', in Y. Bar-Hillel (ed.), *Proceedings of the 1964 International Congress for Logic, Methodology and Philosophy of Science*, 1965a.
Hintikka, Jaakko, 'Towards a Theory of Inductive Generalization', in Y. Bar-Hillel (ed.), *Proceedings of the 1964 International Congress for Logic, Methodology and Philosophy of Science*, 1965b.
Hintikka, Jaakko, 'On a Combined System of Inductive Logic', *Studia Logica-Mathematica et Philosophica, Acta Philosophica Fennica* **18** (1965).
Hintikka, Jaakko, 'Knowledge, Acceptance and Inducative Logic', in J. Hintikka and P. Suppes (eds.), *Aspects of Inductive Logic*, 1966a, pp. 1–20.
Hintikka, Jaakko and Suppes, P. (eds.), *Aspects of Inductive Logic*, North-Holland, Amsterdam, 1966.
Hodges, J. L., *Elements of Finite Probability*, Holden-Day, San Francisco, 1965.
Hodges, J. L. and Lehmann, E. L., *Basic Concepts of Probability and Statistics*, Holden Day, San Francisco, 1964.
Horowitz, Irving L., 'Establishment Sociology: The Value of Being Value-Free', *Inquiry* **6** (1963), 129–39.
Hospers, J. and Sellars, W. (eds.), *Readings in Ethical Theory*, Appleton, 1952.
Houthakker, H. S., 'Revealed Preference and the Utility Function', *Economica* **17** (1950), 159–74.

Houthakker, H. S., 'The Logic of Preference and Choice', in A. T. Tymieniecka, (ed.), *Contributions to Logic and Methodology in Honour of J. M. Bocheński*, Amsterdam 1965, pp. 193–207.
Hull, Clark L., 'Value, Valuation and Natural Science Methodology', *Phil. Sci.* (1944), 125–41.
Hull, Clark L., *A Behaviour System*, Yale University Press, New Haven, 1952.
Humphreys, W. C., 'Statistical Ambiguity and Maximal Specificity', *Phil. Sci.* **35** (1968), 112–5.
Hurst, M. P. and Siegle, S., 'Prediction of Decisions from a Metric Scale of Utility', *J. Experimental Psych.* **52** (1956), 128–34.

Irving, John A., *Science and Values*, Toronto 1952.
Irwin, F. W., 'Stated Expectations as Functions and Desirability of Outcome', *J. Personality* **21** (1953), 329–35.
Irwin, F. W. and Smith, W. A. S., 'Value, Cost and Information, as Determiners of Decision', *J. Exper. Psych.* **54** (1957), 229–32.
Iwand, Thomas, 'Ethical Systems as Order Relations', *Ratio* **7** (1965), 145–63.

Jacobs, Philip, 'Functions of Values in Policy Decisions', *Amer. Behav. Sc.* Supplementary 5, No. 9 (1962).
Jeffrey, Richard C., 'Popper and the Rule of Succession', *Mind* **73** (1964), 129.
Jeffrey, Richard C., 'Ethics and the Logic of Decision', *J. Phil.* **62** (1965a), 528–39.
Jeffrey, Richard C., *The Logic of Decision*, McGraw-Hill, Toronto, 1965b.
Jeffrey, Richard C., 'Valuation and the Acceptance of Scientific Hypotheses', *Phil. Sci.* **23** (1956).
Jeffreys, Sir Harold, *Theory of Probability*, 1939, 3rd ed., Clarendon Press, Oxford, 1961.
Jeffreys, Sir Harold, 'The Present Position in Probability Theory', *Brit. J. Phil. Sci.* **5** (1954–55), 275–89.
Jennings, R. E., 'Reference and Choice as Logical Correlatives', *Mind* **76** (1967), 556–67.
Johnson, Oliver A., 'The Justification of Belief', *Dialogue* **4** (1965), 336–50.

Kadish, Mortimer R., 'Evidence and Decision', *J. Phil.* **68**, (1951), 229–42.
Kahl, R. (ed.), *Studies in Explanation. A Reader in the Philosophy of Science*, Prentice-Hall, 1963.
Katz, Jerrold K., *The Problem of Induction and Its Solution*, University of Chicago Press, Chicago, 1962, reviewed by D. Stove, *Aust. J. Phil.* **41** (1963).
Kaufman, A. S., 'Aims of Scientific Activity', *Monist* **52** (1968), 374–89.
Kaufmann, Felix, 'The Logical Rules of Scientific Procedure', *Phil. Phenomenol. Res.* **2** (1941–42), 457–71.
Kaufmann, Felix, 'Scientific Procedure and Probability', *Phil. Phenomenol. Res.* **6** (1945–46), 47–66.
Keene, G. B., 'Confirmation and Corroboration', *Mind* **70** (1961), 85–7.
Kelly, J. H., 'Entropy of Knowledge', *Phil. Sci.* **36** (1969), 178–96.
Kemble, E. C., 'Is the Frequency Theory of Probability Adequate for All Scientific Purposes?', *Amer. J. of Physics* **10** (1942), 6–16.
Kemeny, J. G., 'A Logical Measure Function', *Journal of Symbolic Logic* **18** (1953), 289–308.
Kemeny, J. G., 'Two Measures of Complexity', *J. Phil.* **52** (1955a), 722–33.

Kemeny, J. G., 'Fair Bets and Inductive Probabilities', *Journal of Symbolic Logic* **20** (1955b), 263-73.
Kemeny, J. G., 'Carnap's Theory of Probability and Induction', in P. A. Schilpp (ed.), *The Philosophy of Rudolf Carnap*, 1963.
Kemeny, J. G. and Snell, J. L., *Mathematical Models in the Social Sciences*, New York 1965.
Kemeny, J. G. and Snell, J. L., 'Preference Rankings: An Axiomatic Approach', in J. G. Kemeny and J. L. Snell (ed.), *Mathematical Models in the Social Sciences*, New York, 1962, pp. 9-23.
Kenny, A. J. P., 'Practical Inference', *Analysis* **26** (1965-66), 65.
Keynes, John Maynard, *A Treatise on Probability*, Macmillan, London, 1921.
Kneale, William, *Probability and Induction*, Clarendon Press, Oxford, 1950.
Kneale, William, 'Probability and Induction', *Mind* **60** (1951), 310-7.
Kneebone, G. T., 'Induction and Probability', *Proc. Arist. Soc.* **50** (1949-50), 27-42.
Kolmogorov, A., *The Foundations of Probability* (transl. by M. Morrison), Chelsea, New York, 1960.
Kolnai, Aurel, 'Games and Aims', *Proc. Arist. Soc.* **66** (1965-66), 103-28.
Koopman, B. O., 'The Axioms and Algebra of Intuitive Probability', *Annals of Mathematics Series 2*, **42** (1940a), 269-92.
Koopman, B. O., 'The Bases of Probability', *Bull. Amer. Math. Soc.* **46** (1940b), 763-74.
Koopman, B. O., 'Intuitive Probabilities and Sequences', *Ann. Math.* **42** (1941), 169-87.
Koopman, B. O., 'Intuitive Probabilities and Series', *Annals of Mathematics Series 2*, **42** (1941), 169-87.
Koopmans, T. C., 'On Flexibility of Future Preferences', in M. W. Shelly and G. L. Bryan, 1964.
Korner, S. (ed.), *Observation and Interpretation*, Butterworths, London.
Krantz, D. H. and Tversky, A., 'A Critique of the Applicability of Cardinal Utility Theory', *Michigan Mathematical Psychology Program 65-4*, University of Michigan, 1965.
Krausser, P., 'Dilthey's Revolution in the Theory of the Structure of Scientific Inquiry and Rational Behaviour', *Rev. Met.* **22** (1968).
Kurtz, Paul W., 'Human Nature, Homeostasis and Value', *Phil. Phenomenol. Res.* **17** (1956-57), 36-55.
Kyburg, Henry E. Jr., 'R. B. Braithwaite on Probability and Induction', *Brit. J. Phil. Sci.* **9** (1958-59), 203-20.
Kyburg, Henry E. Jr., 'Demonstrative Induction', *Phil. Phenomenol. Res.* **21** (1960-61), 80-92.
Kyburg, Henry E. Jr., *Probability and the Logic of Rational Belief*, Wesleyan University Press, Middleton, Connecticut, 1961.
Kyburg, Henry E. Jr., 'A Further Note on Rationality and Consistency', *J. Phil.* **60** (1963), 463-5.
Kyburg, Henry E. Jr., 'Recent Work in Inductive Logic', *Amer. Phil. Quart.* **1** (1964), 249-87.
Kyburg, Henry E. Jr., 'Probability and Decision', *Phil. Sci.* **33** (1966), 250-61.
Kyburg, Henry E. Jr., 'Review of P. A. Schilpp's *The Philosophy of Rudolf Carnap*', *J. Phil.* **65** (1968a), 503-15.
Kyburg, Henry E. Jr., 'Bets and Beliefs', *Amer. Phil. Quart.* **5** (1968b), 54-63.
Kyburg, Henry E. Jr., *Probability Theory*, Prentice-Hall, Englewood Cliffs, New Jersey, 1969.

Kyburg, Henry E. Jr. and Harper, William L., 'The Jones Case', *Brit. J. Phil. Sci* **19** (1968), 247–51.
Kyburg, Henry E. Jr. and Smokler, Howard, *Studies in Subjective Probability*, John Wiley and Sons, New York, 1964.
Lamont, William Dawson, *The Value Judgement*, Edinburgh 1955.
Lange, O., 'The Determinateness of the Utility Function', *Review of Economic Studies* **1** (1934), 218–25.
Lawrie, Norman White, D., and Donaldson, W., *Operational Research Techniques*, Business Books, London, 1969.
Leach, James, 'Historical Objectivity and Value Neutrality', *Inquiry* **11** (1968a), 349–67.
Leach, James, 'Explanation and Value Neutrality', *Brit. J. Phil. Sci.* **19** (1968b), 93–108.
Leblanc, Hugues, 'Two Probability Concepts', *J. Phil.* **53** (1956), 679–88.
Leblanc, Hugues, 'A New Interpretation of $c(h, e)$', *Phil. Phenomenol. Res.* **21** (1960–61), 373–6.
Leblanc, Hugues, *Statistical and Inductive Probabilities*, Prentice-Hall, New York, 1962.
Lehmann, E. L., *Testing Statistical Hypotheses*, John Wiley and Sons, New York, 1959.
Lehman, R. S., 'On Confirmation and Rational Betting', *J. Symbolic Logic* **20** (1955), 251–62.
Lehrer, Keith, 'Letter: On Knowledge and Probability', *J. Phil.* **61** (November 1961).
Lehrer, Keith, 'Decisions and Causes', *Phil. Rev.* **72** (1963), 224–7.
Lehrer, Keith, 'Knowledge and Probability', *J. Phil.* **61** (1964), 368–72.
Lehrer, Keith, 'Letter: On Knowledge and Probability', *J. Phil.* **62** (1965), 67–8.
Lehrer, Keith, 'Belief and Knowledge', *Phil. Rev.* **77** (1968), 491–9.
Lehrer, Keith, Roelofs, R., and Swain, M., 'Reasons and Evidence, An Unsolved Problem', *Ratio* **9** (1967), 38–48.
Leply, Ray, *Verifiability of Value*, Columbia University Press, New York, 1944
Levi, Isaac, 'Corroboration and Rules of Acceptance', *Brit. J. Phil. Sci.* **13** (1963), 307–13.
Levi, Isaac, 'Hacking Salmon on Induction', *J. Phil.* **62** (1965a), 481–7.
Levi, Isaac, 'Deductive Cogency in Inductive Inference', *J. Phil.* **62** (1965b), 68–77.
Levi, Isaac, 'Recent Work in Probability and Induction', *Synthese* **16** (1966a), 234–44.
Levi, Isaac, 'On Potential Surprise', *Ratio* **8** (1966b), 102–29.
Levi, Isaac, 'Probability Kinematics', *Brit. J. Phil. Sci.* **18** (1967a), 197–209.
Levi, Isaac, 'Information and Inference', *Synthese* **17** (1967b), 369–91.
Levi, Isaac, *Gambling with Truth*, Alfred A. Knopf, New York, 1967c.
Levi, Isaac, 'If Jones Only Knew More', *Brit. J. Phil. Sci.* **20** (1968), 153–9.
Levi, Isaac, 'Decision Theory and Confirmation', *J. Phil.* **58** (1961), 614–25.
Levi, Isaac, 'Must the Scientist Make Value Judgements,' *J. Phil.* **67** (1961), 348.
Levi, Isaac, 'On the Seriousness of Mistakes', *Phil. Sci.* **29** (1962), 50ff.
Levi, Isaac and Morgenbesser, Sidney, 'Belief and Disposition', *Amer. Phil. Quart.* **1** (1964), 221–32.
Levy, Sheldon G., *Inferential Statistics in the Behavioural Sciences*, Holt, Rinehart and Winston, 1968.
Lewis, C. I., *An Analysis of Knowledge and Valuation*, 1946, Open Court, 1962.
Lucas, J. R., 'The One Concept of Probability', *Phil. Phenomenol. Res.* **26** (1965–66), 180–201.
Luce, R. D., 'Semiorders and a Theory of Utility Discrimination', *Econometrica* **24** (1956), 178–91.

Luce, R. D., 'A Probabilistic Theory of Utility', *Econometrica* **26** (1958), 193–224.
Luce, R. D., *Individual Choice Behaviour*, John Wiley and Sons, New York, 1959.
Luce, R. D., 'Sufficient Conditions for the Existence of a Finitely Additive Probability Measure', *Annals of Mathematical Statistics* **38** (1967), 780–6.
Luce, R. D. and Raiffa, H., *Games and Decisions*, John Wiley and Sons, New York, 1957.
Luce, R. D. and Suppes, P., 'Preference, Utility and Subjective Probability', in R. D. Luce, R. R. Bush, and E. Galanter (eds.), 1965.
Luce, R. D., Bush, R. R., and Galanter, E., *Handbook of Mathematical Psychology*, Vol. III, John Wiley and Sons, New York, 1965.
Lundberg, George A., 'Science, Scientists and Values', *Social Forces* **30** (1952), 373–9.
Lyons, David, *Forms and Limits of Utilitarianism*, Oxford 1965.

Machol, R. E., *Information and Decision Processes*, McGraw-Hill, New York, 1960.
Mackenzie, J. C., 'Prescriptivism and Rational Behaviour', *Phil. Quart.* **18** (1968), 310–9.
Mackie, J. L., 'The Paradox of Confirmation', *Brit. J. Phil. Sci.* **13** (1963), 265–77.
Mackie, J. L., 'Miller's So-Called Paradox of Information', *Brit. J. Phil. Sci.* **17** (1966–67), 144–7.
Majumdar, T., *The Measurement of Utility*, Macmillan, London, 1958.
Margenau, Henry, 'On the Frequency Theory of Probability', *Phil. Phenomenol. Res.* **6** (1945–46), 11–25.
Margenau, Henry, 'The Scientific Basis of Value Theory', in A. H. Maslow (ed.), *New Knowledge in Human Values*, 1958.
Marshak, J., 'Rational Behaviour, Uncertain Prospects and Measurable Utility', *Econometrica* **18** (1950), 111–41.
Marshak, J., 'Probability in the Social Sciences', in P. Lazarsfeld (ed.), *Mathematical Thinking in the Social Sciences*, Free Press, New York, 1954.
Marshak, J., 'Toward a Preference Scale for Decision-Making', in M. Shubik (ed.), 1954.
Marshak, J. and Radner, R., 'Note on Some Proposed Decision Criteria', in R. M. Thrall, C. H. Coombs, and R. L. Davis (eds.), *Decision Processes*, John Wiley and Sons, New York, 1954.
Marshak, J., 'Actual versus Consistent Decision Behaviour', *Behavioural Science* **9** (1964), 103–10.
Marshak, J. and Davidson, D., 'Experimental Tests of a Stochastic Decision Theory', in C. W. Churchman and P. Ratoosh (eds.), 1959.
Martin, R. M., *Towards a Systematic Pragmatics*, North-Holland, Amsterdam, 1959.
Martin, R. M., *Intension and Decision*, Prentice-Hall, Englewood Cliffs, New Jersey, 1963.
Marx, Melvin (ed.), *Theories in Contemporary Psychology*, MacMillan, New York, 1963).
Massey, Gerald C. J., 'Hempel's Criterion of Maximal Specificity', *Phil. Studies* **19** (1968).
Matthews, Gwynneth, 'A Note on Inference as Action', *Analysis* **16** (1956), 116–7.
May, K., 'Intransitivity, Utility and the Aggregation of Preference Patterns', *Econometrica* **22** (1954), 1–13.
Mayo, W., 'Probability Models and Thought and Learning Processes', *Synthese* **15** (1963), 204–22.

McClintok, C. G. and Becker, G. M., 'Values: Behavioural Decision Theory', *Annual Review of Psychology* **18** (1967), 239–86.
Meckler, Lester, 'The Value Theory of C. I. Lewis', *J. Phil.* **47** (1950), 565–79.
Meehl, P., 'Theory Testing in Psychology and Physics, A Methodological Paradox', *Phil. Sci.* **34** (1967), 103–15.
Mellor, D. H., 'Connectivity, Chance and Ignorance', *Brit. J. Phil. Sci.* **18** (1967), 235–8.
Mellor, W. W., 'Knowing, Believing and Behaving', *Mind* **56** (1967), 327–45.
Michalos, Alex C., 'Estimated Utility and Corroboration', *Brit. J. Phil. Sci.* **16** (1966–67), 327–31.
Michalos, Alex C., 'Postulates of Rational Preference', *Phil. Sci.* **34** (1967), 18–22.
Miller, David, 'A Paradox of Information', *Brit. J. Phil. Sci.* **17** (1966–67a), 59–61.
Miller, David, 'On a So-Called So-Called Paradox', *Brit. J. Phil. Sci.* **17** (1966–67b), 147–9.
Milnor, J., 'Games Against Nature', in R. M. Thrall, C. H. Coombs, and R. L. Davis (eds.), *Decision Processes*, John Wiley and Sons, New York, 1954.
Minas, J. Sayer, 'Comments on Richard C. Jeffrey's "Ethics and the Logic of Decision"', *J. Phil.* **62** (1965), 542–4.
Minas, J. Sayer and Haworth, L., 'Concerning Value Science', *Phil. Sci.* **21** (1954), 54–61.
Mischel, Theodore, 'Pragmatic Aspects of Explanation', *Phil. Sci.* **33** (1966), 40–60.
Moch, F., 'Reflexions sur les Probabilités', *Dialectica* **11** (1957), 375–91.
Morris, Charles W., and Jones, Lyle V., 'Value Scales and Dimensions' (1955), 523–35.
Morris, Charles W., *Varieties of Human Value*, Chicago 1956.
Moser, Shia, 'Decisions, Commands and Moral Judgments', *Phil. Phenomenol. Res.* **18** (1957–58), 471–88.
Moses, L. E. and Chernoff, H., *Elementary Decision Theory*, John Wiley and Sons, New York, 1959.
Mosteller, F. and Nogee, P., 'An Experimental Measure of Utility', *J. Pol. Econ.* **59** (1951), 371–404.
Murphy, A. E., *The Theory of Practical Reason*, LaSalle, 1964.
Myers, Gerald E., 'Justifying Belief Assertions', *J. Phil.* **64** (1967), 210–4.
Myrdal, Gunnar, *Value in Social Theory*, New York 1958.
McAdam, James I., 'Choosing Flippantly or Non-Rational Choice', *Analysis* **25** (Supplementary) (1965) 132–6.
McCowan, R. S., 'Predictive Policies', *Pro. Arist. Soc.* (Supplementary) **41** (1967), 57–76.

Nagel, Ernest, 'The Meaning of Probability', *J. Amer. Stat. Assoc.* **31** (1936), 10–29.
Nagel, Ernest, *Principles of the Theory of Probability*, Inter. Encyclopedia of Unified Sci., University of Chicago Press, Chicago, 1939.
Nagel, Ernest, 'Probability and Non-Demonstrative Inference', *Phil. Phenomenol. Res.* **5** (1944–45), 485–507.
Nagel, Ernest, *The Structure of Science*, Harcourt and Brace, New York, 1961.
Nagel, Ernest, 'Carnap's Theory of Induction', in P. A. Schilpp (ed.) *The Philosophy of Rudolf Carnap*.
Nagel, Ernest, Suppes, P., and Tarski, A., *Logic, Methodology and Philosophy of Science*, Stanford University Press, Stanford, 1962.
Neyman, Jerzy, 'Outline of a Theory of Statistical Estimation Based on the Classical

Theory of Probability', *Philosophical Transactions of the Royal Society of London*, Series A, **236** (1937), 332–80.

Neyman, Jerzy, *Lectures and Conferences on Mathematical Statistics*, Washington 1938.

Neyman, Jerzy, 'Basic Ideas and Some Recent Results of the Theory of Testing Statistical Hypotheses', *Royal Statistical Society* **105** (1942), 292–327.

Neyman, Jerzy, *First Course in Probability and Statistics*, Henry Holt, New York, 1950.

Neyman, Jerzy (ed.), *Proceedings of the Fourth Berkeley Symposium on Mathematical Statistics and Probability*, Vol. 1, University of California Press, Berkeley, California, 1961.

Neyman, Jerzy, '"Inductive Behaviour" as a Basic Concept of Philosophy of Science', *Review of the International Statistical Institute* **25** (1967), 7–22.

Neyman, Jerzy and Pearson, E. S., 'The Testing of Statistical Hypotheses in Relation to Probabilities A Priori', *Pro. Cambridge Phil. Soc.* **29** (1932–33).

Nielsen, H. A., 'Sampling and the Problem of Induction', *Mind* **68** (1959), 474–81.

Nogee, P. and Mosteller, F., 'An Experimental Measure of Utility', *J. Pol. Econ.* **59** (1951), 371–404.

OConnor, J., 'How Decisions are Predicted', *J. Phil.* **64** (1967), 429–30.

Ofstad, Harold, 'Objectivity of Norms and Value-Judgments According to Recent Scandinavian Philosophy', *Phil. Phenomenol. Res.* **13** (1952–53), 42–68.

Oliver, Henry M., Jr., *A Critique of Socio-Economic Goals*, Bloomington 1954.

Oppenheim, Felix E., 'Rational Choice', *J. Phil.* **50** (1953), 341–50.

Parsons, Talcott and Shils, E. A. (ed.), *Toward a General Theory of Action*, Cambridge, Mass., 1951.

Pakswer, S., 'Information, Entropy and Inductive Logic', *Phil. Sci.* **21** (1954), 254–9.

Perry, Ralph Barton, *General Theory of Value*, Longmans, Green and Co., New York, 1926.

Pikler, Andrew G., 'Utility Theories in Field Physics and Mathematical Economics', *Brit. J. Phil. Sci.*, (1954–55), 47–58; 303–18.

Popper, Karl R., 'Two Autonomous Axiom Systems of the Calculus of Probabilities', *Brit. J. Phil. Sci.* **6** (1955–56), 51–7.

Popper, Karl R., 'The Aims of Science', *Ratio* **1** (1957a), 24–35.

Popper, Karl R., 'Probability Magic or Knowledge out of Ignorance', *Dialectica* **11** (1957), 354–74.

Popper, Karl R., 'The Propensity Interpretation of Probability', *Brit. J. Phil. Sci.* **10** (1959), 25–42.

Popper, Karl R., *The Logic of Scientific Discovery*, Hutchinson, London, 1959.

Popper, Karl R., 'Probabilistic Independence and Corroboration by Experimental Test', *Brit. J. Phil. Sci.* **10** (1959–60), 315–8.

Popper, Karl R., 'Creative and Non-Creative Definitions in the Calculus of Probabilities', *Synthese* **15** (1963), 167–86.

Popper, Karl R., *Conjectures and Refutations*, Basic Books, New York, 1962.

Popper, Karl R., 'A Paradox of Zero Information', *Brit. J. Phil. Sci.* **17** (1966–67), 141–3.

Popper, Karl R., 'A Comment on Miller's New Paradox of Information', *Brit. J. Phil. Sci.* **17** (1966–67), 61–9.

Popper, Karl R., 'The Mysteries of Udolfo: A Reply to Professors Jeffrey and Bar-Hillel', *Mind* **76** (1967), 103–10.

Pollack, Robert A., 'Additive von Neumann-Morgenstern Utility Functions', *Econometrica* **35** (1967), 485–94.
Pratt, J. W., Raiffa, H., and Schlaifer, R., 'The Foundations of Decisions Under Uncertainty: An Elementary Exposition', *J. Amer. Stat. Assoc.* **59** (1964), 353–75.
Pratt, J. W., Raiffa, H., and Schlaifer, R., *Introduction to Statistical Decision Theory*, McGraw-Hill, New York, 1965.
Prior, A. N., 'The Runabout Inference Ticket', *Analysis* **21** (1960), 38–9.
Putnam, Hillary, 'Degree of Confirmation and Inductive Logic', in P. A. Schilpp (ed.), *The Philosophy of Rudolf Carnap*.

Quine, W. V., 'The Scope and Language of Science', *Brit. J. Phil. Sci.* **8** (1957–58), 1–17.

Radar, T., 'The Existence of a Utility Function to Represent Preferences', *Rev. of Econ. Studies* **30** (1963), 229–32.
Radcliff, Peter, 'Beliefs, Attitudes and Actions', *Dialogue* **4**, (1966), 456–64.
Raiffa, Howard, *Decision Analysis: Introductory Lectures on Choices Under Uncertainty*, Addison-Wesley Reading, Mass., 1968.
Ramsey, Frank Plumpton, in R. B. Braithwaite (ed.), *The Foundations of Mathematics and Other Logical Essays*, 1931, Routledge and Kegan Paul, London, 1954.
Rapoport, A., 'Lewis F. Richardson's Mathematical Theory of War', *J. of Conflict Resolution* **1** (1957), 244–99.
Rapoport, A., 'Critiques of Game Theory', *Behavioural Science* **4** (1959), 49–66.
Rapoport, A., *Fights, Games and Debates*, University of Michigan Press, Ann Arbor, 1960.
Rapoport, A., 'Mathematical Models of Social Interaction', in R. D. Luce (ed.), *Handbook of Mathematical Psychology*, John Wiley and Sons, New York.
Rapoport, A., 'Game Theory and Human Conflict', in E. B. McNeil (ed.), *The Nature of Human Conflict*, Vol. 10 (1965), pp. 195–226, Prentice-Hall, Englewood Cliffs.
Rapoport, A., *Two-Person Game Theory: The Essential Ideas*, University of Michigan Press, Ann Arbor, 1966.
Rapoport, A. and Orwant, C. J., 'Experimental Games: A Review', *Behavioural Science* **7** (1962), 1–37.
Rapoport, A., Chammah, A., Dwyer, J., and Gyr, J., 'Three-Person Non-Zero-Sum Non-Negotiable Games', *Behavioural Science* **7** (1962), 38–58.
Rapoport, A. and Chammah, A. M., *Prisoner's Dilemma: A Study in Conflict and Cooperation*, University of Michigan Press, Ann Arbor, 1965.
Reichenbach, Hans, *The Theory of Probability*. University of California Press, Berkeley, California., 1949.
Rescher, Nicholas, 'Non-Deductive Rules of Inference and Problems in the Analyses of Inductive Reasoning', *Synthese* **13** (1961).
Rescher, Nicholas, 'The Ethical Dimension of Scientific Research', in R. Colodny (ed.), 1965, pp. 261–76.
Rescher, Nicholas, (ed.) *The Logic of Decision and Action*, University of Pittsburgh Press, 1966a.
Rescher, Nicholas, 'Practical Reasoning and Values', *Phil. Quart.* **16** (1966b), 806–17.
Rescher, Nicholas, 'Values and the Explanation of Behavior', *Phil. Quart.* **17** (1967a), 130–6.
Rescher, Nicholas, 'Semantic Foundations of the Logic of Preference', in N. Rescher (ed.), *The Logic of Decision and Action*, 1967b.

Rescher, Nicholas, *Introduction to Value Theory*, Prentice-Hall, Englewood Cliffs, New Jersey, 1969.
Rescher, Nicholas, *Distributive Justice*, New York 1966a.
Rescher, Nicholas, 'Notes on Preference, Utility and Cost', *Synthese* **16** (1966b), 332–343.
Rescher, Nicholas, 'The Dynamics of Value Change', in K. Bair and N. Rescher (eds.), *Values and The Future*, New York 1968.
Robertson, D. H., *Utility and All That*, George Allen and Unwin, London, 1952.
Robertson, D. H., 'Utility and All What?', *Economic Journal* **64** (1954), 665–78.
Rose, A. M., 'Values in Social Research', in idem, *Theory and Method in the Social Sciences*, Minneapolis 1954.
Rose, A. M., 'Sociology and the Study of Values', *Brit. J. Sociology* **7** (1966), 1–17.
Roshwald, M., 'Value Judgments in the Social Sciences', *Brit. J. Phil. Sci.* **6** (1955–56), 186–208.
Rotenstreich, Nathan, 'The Value Aspect of Science', *Phil. Phenomenol. Res.* **20** (1959–60), 513–20.
Rozeboom, William W., 'New Dimensions of Confirmation', *Phil. Sci.* **35** (1968), 135–55.
Rudner, Richard, *The Philosophy of Social Science*, Prentice Hall, Englewood Cliffs, New Jersey.
Rudner, Richard, 'The Scientist Qua Scientist Makes Value Judgements', *Phil. Sci.* **20** (1953).

Salmon, Wesley C., 'The Status of Prior Probabilities in Statistical Explanation', *Phil. Sci.* **32** (1965), 137–46.
Salmon, Wesley C., *The Foundations of Scientific Inference*, University of Pittsburgh Press, Pittsburgh, 1967.
Salmon, Wesley C., Barker, S. F., and Kyburg, H. E., Jr., 'Symposium: Inductive Evidence', *Amer. Phil. Quart.* **2** (1965), 265–87.
Sasieni, Maurice W. and Ackoff, R. L., *Fundamentals of Operations Research*, John Wiley and Sons, New York, 1968.
Savage, Leonard J. and Friedman, M., 'The Utility Analysis of Choices Involving Risks', *J. Pol. Econ.* **58** (1948) 279–304.
Savage, L. J., 'The Theory of Statistical Decisions', *J. Amer. Statistical Association*, **46** (1951), 55–67.
Savage, L. J., *The Foundations of Statistics*, John Wiley and Sons, New York, 1954.
Savage, L. J., 'Bayesian Statistics', in R. E. Machol and P. Gray (eds.), (1962a).
Savage, L. J., *The Foundations of Statistical Inference*, John Wiley and Sons, New York, 1962b.
Savage, L. J., 'Bayesian Statistical Inference for Psychological Research', *Psych. Rev.* **70** (1963), 193–242.
Savage, L. J., 'Implications of Personal Probability for Induction', *J. Phil.* **64** (1967a), 593–607.
Savage, L. J., 'Difficulties in the Theory of Personal Probability', *Phil. Sci.* **34** (1967), 333–40.
Schackle, George Lennox, *Expectation in Economics*, 2nd ed., Cambridge University Press, 1952.
Schackle, George Lennox, *Decision, Order and Time in Human Affairs*, Cambridge 1961.
Scheffler, Israel, *Anatomy of Inquiry*, Alfred Knopf, New York, 1963.
Schelling, T. C., *The Strategy of Conflict*, Cambridge, Mass., 1960.
Schick, Frederick, 'Consistency and Rationality', *J. Phil.* **60** (1963), 5–19.

Schick, Frederick, 'Review of The Logic of Decision', by R. C. Jeffrey, *J. Phil.* **64** (1967), 396–400.
Schick, Frederick, 'Arrow's Proof and the Logic of Preference', *Phil. Sci.* **36** (1969), 127–44.
Schlaifer, R., *Probability and Statistics for Business Decisions*, McGraw-Hill, New York, 1961.
Schlaifer, R. and Raiffa, H., *Applied Statistical Decision Theory*, Harvard School of Business Administration, Boston, 1961.
Schmidt, P. F., 'Ethical Norms in Scientific Method', *J. Phil.* **56** (1959), 644–52.
Schoeck, H. and Wiggins, J. W. (eds.), *Scientism and Values*, Princeton 1960.
Scriven, Michael, *The Methodology of Evaluation*. Publication 110 of the Social Science Education Consortium, 1966.
Sellars, W., 'Inference and Meaning', *Mind* **62** (1953), 313–38.
Sellars, W., 'Imperatives, Intentions and the Logic of "Ought"', in H. Castañeda and G. Nakhnikian (eds.), *Morality and the Language of Conduct*, Detroit 1963, pp.159–218.
Sellars, W., 'Induction as Vindication', *Phil. Sci.* **31** (1964), 197–231.
Sellars, W., 'Thought and Action', in Keith Lehrer (ed.), *Freedom and Determinism*, Random House, 1966.
Shelly, Maynard Wolfe and Bryan, Glenn L., *Human Judgments and Optimality*, John Wiley and Sons, New York, 1964.
Shimony, Abner, 'Amplifying Personal Probability', *Phil. Sci.* **34** (1967), 314–25.
Shubik, M. (ed.), *Readings in Game Theory and Political Behaviour*, Doubleday, Garden City, New York, 1954.
Shubik, M., *Strategy and Market Structure*, John Wiley and Sons, New York, 1959.
Siegel, Sidney, 'A Method for Obtaining an Ordered Metric Scale', *Psychometricka* **21** (1956), 207–16.
Siegel, Sidney, 'Level of Aspiration and Decision-Making', *Psych. Rev.* **64** (1957), 253–62.
Siegel, Sidney, in Samuel Messik (ed.), *Decision and Choice*, McGraw-Hill, Toronto, 1964.
Silander, F. S. and Wassenman, P., *Decision-Making: An Annotated Bibliography*, Cornell University, Graduate School of Business and Public Administration, 1958.
Simon, H. A., 'A Behavioral Model of Rational Choice', *Quart. J. Econ.* **69** (1955), 99–118.
Simon, H. A., *Models of Man, Social and Rational*, John Wiley and Sons, New York, 1957.
Simon, H. A., 'The Logic of Rational Decision', *Brit. J. Phil. Sci.* **16** (1965), 169–86.
Simon, H. A., 'The Logic of Heuristic Decision Making', in N. Rescher (ed.), *The Logic of Decision and Action*, 1967, pp. 1–20.
Simon, H. A., *Administrative Behavior: A Study of Decision-Making Processes in Administrative Organization*, 2nd ed., New York 1957.
Simpson, George, 'Science as Morality', *Phil. Sci* **18** (1951), 132–43.
Singer, E. A., Jr., *On the Contented Life*, Henry Holt, New York, 1936.
Singer, E. A., Jr., *In Search of a Way of Life*, Columbia University Press, New York, 1948.
Sleigh, R. C., 'A Note on Some Epistemic Principles of Chisolm and Martin', *J. Phil.* **61** (1964a), 216–8.
Sleigh, R. C., 'A Note on Knowledge and Probability', *J. Phil.* **61** (1964b), 478.

Sloman, Aaron, 'Predictive Policies', *Proceedings of the Aristotelian Society* (Suppl.) **41** (1967), 77–94.
Smith, Nicholas M., Jr., 'A Calculus for Ethics; A Theory of the Structure of Value', *Behavioural Science* **1** (1956), 111–42; 186–211.
Smith, Nicholas M., Jr., 'A Critique of Churchman's Science of Values', *Technical Report TP*-40, McLean, Research Analysis Corp., Virginia, 1961.
Smith, N. M., Walters, S. S., Brooks, F. C., and Blackwell, D. H., 'The Theory of Value and the Science of Decision – A Summary', *J. of the Operations Research Society of America* **1** (1953), 103–13.
Smokler, Howard, 'Goodman's Paradox and the Problem of Rules of Acceptance', *Amer. Phil. Quart.* **3** (1966), 71–6.
Sneed, Joseph D., 'Strategy and the Logic of Decision', *Synthese* **16** (1966), 270–83.
Sneed, Joseph D., 'Entropy, Information and Decision', *Synthese* **17** (1967) 302–407.
Snell, J. L. and Kemeny, J. G., *Mathematical Models in the Social Sciences*, New York 1965.
Somerville, J., 'The Value Problem and Marxist Social Theory', *J. Value Inquiry* **2** (1968), 52–7.
Stevens, S. S., 'Measurement, Psychophysics and Utility', in C. W. Churchman and P. Ratoosh (eds.), 1959.
Stevenson, J. T., 'Roundabout the Runabout Inference Ticket', *Analysis* **21** (1961), 124–8.
Stigler, G., 'The Development of Utility Theory Parts I and II', *J. Pol. Econ.* **58** (1950), 307–27; 373–96.
Stocker, M., 'Knowledge, Causation and Decision', *Noûs* **2** (1968), 65–73.
Strotz, R. H., 'Cardinal Utility', *Amer. Econ. Rev.* **43** (1953), 384–97.
Suppes, P., 'Some Open Problems in the Foundations of Subjective Probability', in R.E. Machol (ed.), 1960.
Suppes, P., 'The Philosophical Relevance of Decision Theory', *J. Phil.* **58** (1961), 605–14.
Suppes, P., 'The Desirability of Formalization in Science', *J. Phil.* **65** (1956), 651–64.
Suppes, P., 'Concept Formation and Bayesian Decisions'.
Suppes, P., 'Probabilistic Inference and the Concept of Total Evidence', in J. Hintikka and P. Suppes (eds.), pp. 21–48.
Suppes, P., 'A Bayesian Approach to the Paradoxes of Confirmation', 198–207, in Hintikka and Suppes (eds.), *Aspects of Inductive Logic*, 1966, pp. 198–207.
Suppes, P., 'Behavioristic Foundations of Utility', *Econometrica* **29** (1961), 186–202.
Suppes, P., Davidson, D., and Seigel, S., 'Some Experiments and Related Theory on the Measurement of Utility and Subjective Probability', in *Applied Mathematics and Statistical Laboratory Technical Report* #1, Stanford University, 1955.
Suppes, P. and Winet, M., 'An Axiomatization of Utility Based on the Notion of Utility Differences', *Management Science* **1** (1955), 259–70.
Suppes, P. and Walsh, K., 'A Non-Linear Model for the Experimental Measurement of Utility', *Behavioural Science* **4** (1959), 204–11.
Suppes, P. and Luce, R. D., 'Preference, Utility and Subjective Probability', in R. D. Luce, R. R. Bush, and E. Galanter (eds.), 1965.

Taguiri, Renato, 'Value Orientations and the Relationships of Managers and Scientists', *Administrative Science Quarterly* **9** (1965), 199–210.
Thorndike, Edward C., 'Science and Values', *Science* **83** (1936), 1–8.

Timur, M., 'Better as the Value-Fundamental', *Mind* **64** (1955), 52–60.
Todd, W., 'Probability and the Theorem of Confirmation', *Mind* **76** (1967), 260–3.
Tversky, A. and Krantz, D. H., 'A Critique of the Applicability of Cardinal Utility Theory', *Michigan Mathematical Psychology Program 65-4*, University of Michigan, 1965.

Urban, Wilbur M., 'Science and Value', *Ethics* **51** (1941), 291–306.
Urmson, J. O., 'Two of the Senses of "Probable"', *Analysis* **8** (1947), 9–16.

Vajda, S., *Theory of Games and Linear Programming*, John Wiley and Sons, New York, 1956.
Van Dantzig, D., 'Carnap's Foundation of Probability Theory', *Synthese* **8** (1950–51), 459–70.
Van Dantzig, D., 'Statistical Priesthood' (Savage on Personal Probabilities), *Statistica Neerlandica* **2** (1957), 1–16.
Venn, John, *The Logic of Chance* (1888), reprinted Chelsea, New York, 1963.
Vickers, John M., 'Some Remarks on Coherence and Subjective Probability', *Phil. Sci.* **32** (1965), 32–8.
Von Mering, Otto, *A Grammar of Human Values*, University of Pennsylvania Press, Pittsburgh, 1961.
Von Mises, Richard, *Probability, Statistics and Truth*, 1928, 2nd German ed. 1936, 2nd revised English ed. by Hilda Geiringer, Macmillan, New York, 1957.
Von Neumann, John and Morgenstern, O., *Theory of Games and Economic Behaviour*, 1947; Princeton 1963.
Von Wright, Georg Henrik, *The Logical Problem of Induction*, Macmillan, New York, 1957.
Von Wright, Georg Henrik, *Norm and Action; A Logical Inquiry*, Humanities Press, New York, 1963a.
Von Wright, George Henrik, 'Practical Inference', *Phil. Rev.* **72** (1963b), 159–79.
Von Wright, Georg Henrik, *The Logic of Preference*, Edinburgh 1963c.
Von Wright, Georg Henrik, *A Treatise on Induction and Probability*, Blackwell, London, 1965.
Von Wright, Georg Henrik, 'The Logic of Action...', in N. Rescher (ed.), *The Logic of Decision and Action*, 1966.
Von Wright, Georg Henrik, 'Deontic Logics', *Amer. Phil. Quart.* **4** (1967), 136–43.
Von Wright, Georg Henrik, *The Varieties of Goodness*, London 1963.

Wald, A., *On the Principles of Statistical Inference*, University of Notre Dame Press, South Bend, Indiana, 1942.
Wald, A., 'Statistical Decision Functions', *Annals of Mathematical Statistics* **20** (1949), 165–205.
Wald, A., *Statistical Decision Functions*, John Wiley and Sons, New York, 1950.
Wedberg, Anders, 'Decision and Belief in Science', *Danish Yearbook of Philosophy*, **1** (1964), 139–58.
Weiss, L., *Statistical Decision Theory*, McGraw-Hill, New York, 1961.
Weisskopf, W. A., 'Hidden Value Conflicts in Economic Thought', *Ethics* **61** (1961), 195–204.
Welty, Gordon, 'Quality Control, Welfare Economics and Prof. Bier', *J. Value Inquiry* **1** (1967), 139–48.

Weyland, F., "A Note on 'Knowledge, Certainty and Probability'", *Inquiry* **7** (1964), 417.
Wigner, Eugene (ed.) *Physical Science and Human Values*, Princeton 1947.
Will, Frederick L., 'The Preferability of Probable Belief', *J. Phil.* **62** (1965), 57–67.
Williams, Donald, 'The Problem of Probability', *Phil. Phenomenol. Res.* **6** (1945–46), 619–22.
Williams, Donald, *The Ground of Induction*, Harvard University Press, Cambridge, Mass., 1947.
Williams, Donald, 'Professor Carnap's Philosophy of Probability', *Phil. Phenomenol. Res.* **13** (1952–53), 103–21.
Williams, Donald, 'On the Direct Probability of Induction', *Mind* **62** (1953), 465–83.
Williams, J. D., *The Complete Strategist*, McGraw-Hill, New York, 1954.
Wisdom, J. O., *Foundations of Inference in Natural Science*, Methuen, London, 1952.